T0310032

WiMedia UWB

WiMedia UWB
TECHNOLOGY OF CHOICE FOR WIRELESS USB AND BLUETOOTH

Ghobad Heidari
Olympus Communication Technology of America, USA

A John Wiley and Sons, Ltd, Publication

This edition first published 2008.
© 2008 John Wiley & Sons, Ltd.

Registered office
John Wiley & Sons Ltd, The Atrium, Southern Gate, Chichester, West Sussex, PO19 8SQ, United Kingdom

For details of our global editorial offices, for customer services and for information about how to apply for permission to reuse the copyright material in this book please see our website at www.wiley.com.

The right of the author to be identified as the author of this work has been asserted in accordance with the Copyright, Designs and Patents Act 1988.

All rights reserved. No part of this publication may be reproduced, stored in a retrieval system, or transmitted, in any form or by any means, electronic, mechanical, photocopying, recording or otherwise, except as permitted by the UK Copyright, Designs and Patents Act 1988, without the prior permission of the publisher.

Wiley also publishes its books in a variety of electronic formats. Some content that appears in print may not be available in electronic books.

Designations used by companies to distinguish their products are often claimed as trademarks. All brand names and product names used in this book are trade names, service marks, trademarks or registered trademarks of their respective owners. The publisher is not associated with any product or vendor mentioned in this book. This publication is designed to provide accurate and authoritative information in regard to the subject matter covered. It is sold on the understanding that the publisher is not engaged in rendering professional services. If professional advice or other expert assistance is required, the services of a competent professional should be sought.

A catalogue record for this book is available from the British Library.

ISBN 978-0-470-51834-2 (H/B)

Set in 11/14pt Times by Aptara Inc., New Delhi, India.
Printed in Great Britain by CPI Antony Rowe, Chippenham, Wiltshire.

To my dear wife,
Constance,
who opened my eyes

Contents

List of Figures

List of Tables

About the Authors

Ghobad Heidari, PhD

Up until January 2008, Dr Heidari was the Vice President of Engineering at Olympus Communication Technology of America, a wholly owned subsidiary of Olympus Corporation. There, he mentored and led a team of multidisciplinary engineers towards the development of a WiMedia UWB chip from concept to tape-out. As a hands-on manager, he individually contributed to baseband design and testing, MAC architectural design, system engineering and integration, as well as the standards. His heavy involvement in the WiMedia Alliance included the co-authorship of the PHY, the MAC, and the WLP specifications of the WiMedia UWB.

Prior to Olympus, Dr Heidari spent several years at Quicksilver Technology, as a Director of Engineering, where he led a team of system engineers in the development of wireless systems on a reconfigurable hardware architecture and the associated software platform. The main applications of interest were in the areas of WWAN and WLAN: cdma2000, WCDMA, and 802.11a.

Dr Heidari also spent several years at Nokia Mobile Phones, designing and developing CDMA-based mobile communication modems (IS-95, cdma2000), and contributing heavily to the associated standards bodies and special interest groups (TR45.5, CDG, 3GPP2).

In the earlier years of his career Dr Heidari worked on the mobile network design, deployment, and optimization, from analytical studies to field tests. In fact, as the Assistant Director of Advanced Technology for a regional baby Bell company, he helped design and deploy the first CDMA cellular network in Chicago.

Dr Heidari received his BSEE from Michigan State University and his MSEE and PhD from Purdue University. While at Purdue, he invented "chaotic communications", a concept based on the use of chaotic systems in digital wireless communications. He has been a part-time faculty member of UCSD Extension since 1999 and served as the chairman of the San Diego chapter of the IEEE Vehicular Technology Society for two years. Dr Heidari has over 30 patents granted or pending. He can be reached at heidari@ieee.org.

Robert T. Short, PhD

PhaseLocked Systems

Dr Short has published a number of seminal papers in the *IEEE Transactions* and other refereed publications. He holds many patents in the wireless area and is an active and successful consultant in the field of wireless systems. Prior to becoming a consultant, Dr Short worked for a number of wireless firms, including Alereon, Sperry Wireless Systems (now a division of L3-Com), and Motorola. Dr Short has founded or co-founded a number of companies and was a professor at the University of Utah. Dr Short is the contributor of Chapter 3 of this book.

Preface

Picture, for a moment, that you have a mobile phone capable of communicating with your computer at such speeds that you can upload movies and music to it within a tenth of the time you are used to over WiFi. Better yet, imagine that you do the same with the external hard drive you have connected to your computer, but without even turning on your computer. Then you get into your car and you have the uploaded movie streamed to the video display in your car for your kids to watch on the trip to grandma's home.

Next, vizualize that you walk into your office and set your laptop on your desk without physically connecting to any docking station or cables. Yet, the laptop is wirelessly communicating with all your peripherals, especially the hard disk and the monitor, as if it were in its docking station. The link speed and quality of service (QoS) over the wireless connection is so high that you feel no degradation in video quality on your monitor or transfer speed to/from your hard drive compared with the wired connections you are used to. Then you take your laptop to a conference room to attend a meeting. Your laptop finds and connects to the projector, the flat-screen display on the wall, as well as to the printer in that room. Moreover, your laptop recognizes and connects to all your colleagues' laptops and smart phones in that room, enabling you to have an ad hoc sharing of ideas and files. In the same conference room, you have a participant who uses her smart phone to stream a video presentation to the projector/display on the wall, as well as to each laptop in the room. The most interesting part of this scenario is not the fact that the wireless links are running at hundreds of Mbps, but that the networking takes place without any preplanning and without the assistance of any intermediary hosts/master devices (computer, access point, etc.).

Then, envisage you are in a hotel room and are using your mobile phone to receive your favorite mobile TV channel. You then create a high-speed connection to the TV/monitor in the room to stream the incoming TV program onto the larger screen for easy viewing. Once you are done watching, you challenge your daughter to a video game you had purchased earlier from a wireless kiosk in the lobby of the hotel. She has her own portable game console, but does not have a copy of the same game on that console. No problem. The two of you wirelessly connect your devices together and to the TV such that you are both playing the game on your phone and watch the game unfold on the large screen. (Of course, you reconfirm to yourself that you are still no match for your teenage daughter at video games).

Now imagine that all of the above scenarios are enabled using a single wireless technology, implemented on a single chip!

As we shall see in this book, WiMedia ultra wideband (UWB) has the technical potential to enable these and tens, if not hundreds, more such short-range high-speed connectivity and networking scenarios. In the age of YouTube and iPhones, such scenarios will increase ever more rapidly in the coming years. As a radio platform for high-speed wireless connectivity and networking, WiMedia UWB is specifically designed to address such applications. Its implementation into a single chip (RF, Baseband, MAC, and protocols) can be practically small, low cost, and of sufficiently low power to be ubiquitously embedded in all sorts of computing, consumer electronic, and mobile devices. In the area of Wireless Personal Area Networks (WPANs), UWB, and specifically WiMedia UWB, is the new (and currently the only) standard technology that is capable of offering a high-speed ad hoc, peer-to-peer networking capability that is keen on power conservation, security, QoS, interoperability and performance.

The Physical (PHY) layer of the WiMedia UWB is currently able to deliver up to 480 Mbps maximum link speeds. There are also plans to increase the speed to more than 1 Gbps in the near future. The output power of the PHY is so low that it is practically at or below the noise level of any narrowband receiver. This helps both in power conservation and in interference reduction.

The WiMedia Medium Access Control (MAC) sublayer complements the PHY layer beautifully, by allowing true peer-to-peer networking with very low overhead and high power efficiency, and by offering a high level of security. More importantly, it does so while providing guaranteed QoS to upper layers. As a result, this MAC protocol is revolutionary among its peers.

The advantage of a standardized UWB technology is that it can be a platform to upgrade many of the currently existing short-range connectivity technologies. For example, Universal Serial Bus (USB) can now have a wireless counterpart, called Certified Wireless USB (CW-USB), based on the WiMedia UWB platform (PHY and MAC). Bluetooth protocol stack can also piggyback on the UWB platform to create a high-speed mode of operation. In fact, the special interest groups (SIGs) for both these technologies (USB-IF and Bluetooth) have already decided to do just that. This way, existing USB and Bluetooth applications, profiles, and protocol stacks may be reusable in a high-speed wireless mode. More importantly, the various upper layer protocols that use the UWB platform are then able to coordinate their activities at the MAC level so as not to interfere with each other.

The importance of this last statement becomes clearer if we recall that peaceful coexistence of disparate wireless technologies in the same radio-frequency (RF) spectrum is not typically achieved. For instance, currently, Bluetooth 2.1 and IEEE 802.11b/g (WiFi) interfere with each other since there is no coordination possible between the two technologies, which use the same frequency band.

Of course, devising a wireless technology that can provide great capabilities is one thing, allocating the RF spectrum to do them in is another. For a high-speed WPAN technology to flourish, it must be able to operate globally. These days, it is inconceivable to require a user to limit the use of their portable WPAN devices to only one region of the world. The world is a much smaller place than a decade ago. Business and leisure travel are apt to take us, along with our electronic gadgets, to various regions on the planet.

Clearly, the user demand for global operation can be met in two ways. First, and the old fashion way, is by implementing extra capabilities (usually in terms of multiple or multi-mode RF frontends) in the device. Second, is by way of worldwide harmonization of the RF spectrum allocation among the various regulatory bodies. The latter is easier said than done, of course. Nonetheless, that is exactly what is being pursued by the UWB industry as the favored approach. We will examine, in the course of this book, how this effort is coming about and what the prospects of such a universal harmonization are.

Having become the de facto standard, WiMedia UWB is destined to capture the WPAN market in personal computing, consumer electronics, and portable/mobile communications products. This may take some time, though, since an ecosystem of WiMedia-enabled devices needs to grow to maturity to allow the full potential of UWB to be realized. This is similar to what Bluetooth and WiFi technologies had to go through before they became prolific. In the meantime, many manufacturers are getting ready for the market by implementing various WiMedia UWB solutions in ever-smaller application-specific integrated circuits.

In this book, I endeavor to demystify the WiMedia UWB technology in many aspects. I have two major goals to achieve. First, to give sufficient background and information related to the WiMedia UWB to make the big picture of this technology highly resolved in the mind of the reader. This includes covering topics from UWB history and applications to worldwide regulations and implementation issues. Second, to provide ample depth of coverage on the technology itself as defined in the standard (ECMA-368). This latter goal is important, since standards are, in general, written in a very cryptic fashion, and ECMA-368 is no exception.

I intended to offer the technical information without requiring a prerequisite on topics such as Orthogonal Frequency Division Multiplexing, MAC, networking, security, etc. Hence, most of the effort was spent in demystifying the standard, both at the PHY and the MAC levels, in such a way that the reader can effortlessly follow the information flow. The PHY and the MAC features are explained at two levels: high level and in depth. The former provides a quick summary of the features and capabilities without delving into much technical complexity. The latter presents an interpretive treatment of the important subjects of the specifications without getting lost in the maze. To tie it all together, I also included a somewhat detailed treatment of the CW-UWB specification and how it uses the WiMedia MAC constructs.

Overall, I hope engineers/researchers, and their managers who intend to implement, integrate, or study WiMedia UWB, CW-USB, High Speed Bluetooth, or WiMedia Logical Link Control Protocol (also known as IP over UWB) will find this book useful. The extensive introductory and regulatory information given in the book should also be of interest to marketing and executive personnel who need to see the broad picture without having to delve into too much detail.

I am grateful to all who have helped me put this work together. Without their help in contributing to or reviewing different sections of this book, it would not have happened. I would especially like to thank Dr. Robert Short for his accepting to join late in the game, and with little advance notice, to contribute the PHY chapter when I was running late with my deadline. Frankly, I think he has done a better job at it than I could have done on my own. I am also indebted to Dr. Alaa Muqattash for his invaluable and detailed review of the MAC and the PAL chapters, and to Dr. Alireza Mehrnia and Christian Politano for their thorough review of the Worldwide Regulations chapter would also like to show my appreciation to Dr. Mahmoud Zadeh for his review of the introductory chapter.

Ghobad Heidari
San Diego

1

Introduction

Ultra wideband (UWB) technology has been contemplated in the past in many different forms. However, only recently has it established a momentum within the industry to become a commercial technology. With the worldwide regulatory bodies, in the process of allowing such a wideband technology to be deployed in an unlicensed frequency band, the final hurdles blocking the proliferation of UWB are being lifted.

The word *wideband* is, of course, a relative term. Until recently, the most wideband commercial signals used in wireless communications were on the order of 20–40 MHz wide. For instance, the term wideband in Wideband Code Division Multiple Access (WCDMA; a third-generation (3G) mobile phone technology) refers to only 5–20 MHz signal bandwidths. On the other hand, UWB signals require bandwidths substantially larger—on the order of 500 MHz. As a result, the signal power density (power per megahertz transmitted) can be minuscule, practically down in the noise level of most other technologies. This makes for a very useful and much needed technology to address two types of user applications:

1. short range, very high data rate (480 Mbps and higher), but at very low energy per transmitted bit;
2. longer range, but much lower data rate, with extremely long battery life.

The latter was pursued and successfully standardized in IEEE 802.15.4a. The former is the focus of this book.

The physical form of the UWB signals has varied quite a bit over time. From pulsed modulation in the earlier days, to spread spectrum and orthogonal frequency modulations today. In fact, engineers have long been experimenting with

WiMedia UWB: Technology of Choice for Wireless USB and Bluetooth   Ghobad Heidari
© 2008 John Wiley & Sons, Ltd

generating and detecting extremely wideband signals as a way of finding novel ways of communication. They have been driven by the promise of low-power signals so widespread in their bandwidth that they would essentially look like noise to any conventional receiver.

Our focus, in this book, will be on ECMA-368 [1], which contains both the Physical Layer (PHY) and the Medium Access Control (MAC) sublayer specifications of WiMedia Alliance (WiMedia for short), a UWB special-interest group (SIG), promoting a Multi-band Orthogonal Frequency Division Multiplexing (MB-OFDM) UWB technology. In addition to the PHY and MAC, this book will discuss the Protocol Adaptation Layers (PALs) such as Certified Wireless Universal Serial Bus (CW-USB), WiMedia Logical Link Control Protocol (WLP) and High Rate Bluetooth.

The rest of this chapter will give a background on UWB technologies in general, and more specifically the work that has been accomplished in recent years. It will build a case for UWB in terms of both technology and commercial market. The standardization effort over the last few years is also summarized, especially the work that is accomplished in WiMedia Alliance. A summary list of features and capabilities of WiMedia UWB is also presented. Finally, some background and terminology elaboration regarding the WiMedia standards will be made to aid the reader with the comprehension of the WiMedia PHY and MAC standard language.

Chapter 2 is devoted to the regulations revolving the use of UWB frequencies around the globe. It focuses on the major market regions of the USA, Japan, Korea, Europe, and China. For each of these regions/countries the UWB regulations are summarized and compared with each other to give a global view of the UWB spectra. This view is very critical in developing a global market for UWB.

Chapters 3 and 4 and are dedicated to the PHY and MAC specifications from WiMedia Alliance respectively. This SIG is behind the publicly available specification ECMA-368 [1], which is published by Ecma International, an international standards organization. Unfortunately, as most standards go, this specification is not written in a reader-friendly fashion, nor is much background information given to allow the reader to understand the reasoning behind its requirements. The flow of information in the standard is not linear either, especially in the MAC sublayer, requiring the reader to jump around the text to find all the necessary requirements and ramifications on a particular topic. These shortcomings are usually expected in standards documents to some extent because

1. they are written for interoperability purposes, not educational purposes;
2. they are written over the course of many months and possibly years;
3. they are the work of many engineers, many of whom contribute to only parts of the overall document, leaving the writing style somewhat inconsistent;

4. they are revised many times in patchwork fashion, as certain fixes/updates require modifications to many sections;
5. the underlying technologies are very complex and little effort is spent in making the documentation easy to read.

Chapters 3 and 4 are designed to remove these shortcomings from ECMA-368. The chapters are by no means a replacement for the standard document, but rather a companion that can overcome the inherent difficulties of understanding the key concepts and requirements embedded in the standard. The descriptive format of these chapters is designed to be educational – far from the style used in the standard. As such, emphasis is put on offering the reader an intuitive understanding of the requirements.

To round out the study of WiMedia UWB, Chapter 5 gives a high-level discussion on PALs being designed/discussed for WiMedia UWB as a platform. The main focus of this chapter is on CW-USB, however. Here, we will review this protocol in comparison with the wired Universal Serial Bus (USB) protocol and with the context of the WiMedia MAC. The frame formats, transaction groups, data flow, connection process, device types, and association model are all discussed. At the end, we offer some thoughts on implementation and interface issues to round out the review of the CW-USB protocol.

1.1 What is UWB?

The history of UWB signals goes back to the very first radios. Using a spark-gap technique, the first radios were, in fact, using ultra-wide bandwidth signals. However, one could argue that those radios were unintentionally UWB. Over the course of the last half a century, the UWB signals have re-emerged.

The history of 'intentional' UWB goes back to 1960s, when pioneering work in the area of impulse generation and reception was being accomplished.[1] At the beginning, UWB stood for the idea that, instead of using conventional transceivers – where a sinusoidal carrier is modulated in amplitude, phase, or frequency to carry information – pulses of very short durations (termed impulses) are transmitted. Owing to the short (nanosecond) duration of the pulses, the bandwidth of such transmissions is much larger than any conventional communication signal. Hence, the terms impulse radio, carrierless technology, and baseband signals were coined. These terms are no longer appropriate, however, since many UWB technologies today are not impulse based or carrierless.

---

[1] UWB has been studied and used in the US Defense Department and other research laboratories in many forms and under many different names. Ross and Robbins patented (1972–1987) the initial applications of UWB signals in areas such as communications and radar [2].

However innovative and unconventional the early impulse radio systems were, the UWB technology has changed dramatically since then. The ultra-wide bandwidth of the signal transmitted is still necessary for it to be considered UWB, but the signal's form is no longer limited to impulses. In fact, these days you can find Direct-Sequence Spread-Spectrum (DSSS) and Orthogonal Frequency Division Multiplexing (OFDM) signals being used in a variety of UWB technologies. Thus, the main difference seemingly left between a general UWB signal and a conventionally 'narrowband' one is the size of the bandwidth. However, there is more to UWB's appeal in recent days than just its signal bandwidth.

In its landmark ruling in 2002, the Federal Communications Commission (FCC) of the United States defined a UWB signal as one that either

1. occupies at least 500 MHz of spectrum,[2] or
2. has a 10 dB bandwidth[3] greater than or equal to 20% of its center frequency.

With this definition, the FCC opened the door for any type of signal and modulation scheme to be admitted as UWB. This begs the question, then, about what makes UWB so special. We have always seen signals grow in bandwidth over the course of the history of communication. There was a time when signal bandwidths were limited to tens of kilohertz. Later, during the Code Division Multiple Access (CDMA) boom days, 1.25 MHz was the top signal bandwidth for any commercial wireless technology. Now, 5–40 MHz signals are easily found in Wireless Wide Area Networks (WWANs) such as WCDMA, as well as Wireless Local Area Networks (WLANs). What is so novel in taking another step in that same direction with UWB?

Obviously, going from 40 MHz to 500 MHz in bandwidth is an extraordinary leap, but the main novelty of the current-day UWB signals can be attributed to two main features:

1. *Overlay usage of the frequency spectrum.* Up until the FCC ruling, the regulatory bodies worldwide followed the usual business of exclusively licensing frequency bands. That is, a license holder has always been the exclusive user of a licensed spectrum. However, with the FCC's historical UWB ruling, this trend was broken for the first time. Now, in the United States, in the 3.1–10.6 GHz

---

[2] The FCC may reduce this threshold to 450 MHz in the future to coordinate with the other regulatory bodies around the world. See Chapter 2 for more details.

[3] A 10 dB bandwidth is the frequency range between the points on either side of the spectral density curve that are 10 dB below the maximum power density point of the curve.

frequency range, the unlicensed UWB users may operate in the same bands in which licensed and other unlicensed operations may be taking place at the same time. In effect, the UWB spectrum allocation has been overlaid on top of the other allocations in that range. Such spectrum sharing would allow more efficient use of the scarce radio spectrum resource.

2. *Operating transmit power below noise floor.* Owing to their large bandwidth, UWB signals can operate at such low spectral density levels that they would theoretically be rendered unrecognizable from noise, from the perspective of a narrowband receiver. Because of this advantage, the FCC was unconcerned about overlaying the UWB spectrum on the radio-frequency (RF) spectrum that is already licensed to other technologies.

Ever since the FCC's ruling on UWB, many new UWB technology areas have been developed, from high data-rate Wireless Personal Area Networks (WPANs) to low data-rate monitoring and control networks. In fact, the FCC categorized the commercial applications of UWB as follows:

1. *Imaging systems (including ground-penetrating radars, wall/through-wall/ medical imaging, and surveillance).* These types of UWB system specialize in imaging applications. Ground-penetrating radar is one such application in which objects or patterns in the ground are detected using UWB pulsed signals. Imaging objects in or behind a nonmetallic wall is also a related application. UWB surveillance systems, although not necessarily an imaging application, use a similar approach in that they monitor an RF perimeter for any intrusions.

2. *Vehicular radar systems.* These systems are entirely related to terrestrial transportation vehicles and are operated in the 24 GHz band. Applications revolve around the safety and handling of vehicles and include collision avoidance, airbag activation, suspension systems, cruise control, etc.

3. *Communication and measurement systems.* These include the familiar area of high data rate, short-range, wireless communication and networking, such as high-speed file transfers and printing, high-definition audio/video streaming, and a myriad of other applications in the consumer electronics (CE), personal computing (PC), and mobile communication arenas. Low data rate, high-precision ranging, location services, and equipment tracking are also added to this category.

This book focuses on the third category of UWB use cases, specifically in the high data-rate arena. Table 1.1 lists some of the comparative parameters and attributes of several technologies and compares them with the UWB technology of

Table 1.1 Comparison of several technologies with WiMedia UWB

Standard	SIG	Max. data rate (Mbps)	Max. range (m)[a]	Frequency band (GHz)	Bandwidth	Attributes
IEEE 802.15.1	Bluetooth	3 (PHY)	10 (class 2)	2.4	1 MHz	Low power; low data rate; short range
IEEE 802.15.4a	ZigBee	1 (PHY)	300	3.1–10.6 (USA)	5 MHz	Precision ranging; low power; robust; portable
IEEE 802.11a	WiFi	54 (PHY)	100	5	20 MHz	High power consumption; medium data rate; no QoS
IEEE 802.11b	WiFi	11 (PHY)	100	2.4	22 MHz	High power consumption; low data rate; no QoS
IEEE 802.11g	WiFi	54 (PHY)	100	2.4	20 MHz	High power consumption; medium data rate; no QoS
IEEE 802.11n	WiFi	100 (MAC)	100	2.4	22 MHz	High power consumption; low data rate; QoS
ECMA-368	WiMedia Alliance	480 (PHY)	10 m	3.1–10.6 (USA)	1.5 GHz	Low power consumption; high data rate; short range; high QoS

[a]Maximum range is achieved at the lowest data rate and usually under ideal environmential/channel conditions.

ECMA-368 [1], which is the main subject of this book. The data rates given in this table are either at the PHY or the MAC sublayer, as indicated.

1.2 Why UWB?

Now, let us explore the key attributes of UWB systems that make them suitable for high-speed WPANs.

1.2.1 Higher Channel Capacity and Power Efficiency

Theoretically, the best way of achieving higher data rates in a communication system is through the increase of the signal bandwidth. Shannon's capacity limit theorem makes this clear for us. Given an additive white Gaussian noise (AWGN) channel, the theorem predicts that the maximum possible error-free information data rate that can be achieved is given by

$$C = B \log_2 \left(1 + \frac{S}{N}\right)$$

where C is the channel capacity in bits/s, B is the signal bandwidth, and S/N is the signal-to-noise ratio in that bandwidth. That is, the channel capacity is directly proportional to signal bandwidth, but only logarithmically proportional to S/N. Thus, the most effective way of increasing the data rate is by increasing the bandwidth, making UWB signals the most efficient in achieving the highest data rate while maintaining a very low signal power density. Put another way, compared with narrowband (or more precisely, narrower-band) signals[4], high-order modulation schemes (which usually require high S/N) are not necessary in UWB signals to achieve very high data rates. Thus, by virtue of its bandwidth advantage over narrowband technologies, UWB is theoretically able to achieve the required data rate at much lower transmission power levels than any narrowband technology.

As a concrete example of the power advantage of UWB systems, let us consider the WiMedia UWB technology. Early indications are that the WiMedia UWB chip designs are able to achieve power consumptions as low as a few hundred milliwatts on average in the active mode. This compares with roughly the same or somewhat more power consumption for WiFi (IEEE 802.11a/b/g). Of course, the exact numbers are highly implementation dependent; but even if we assume that

---

[4] Keep in mind that what is called narrowband in this book may be called wideband in other contexts. For example, the cellular technology WCDMA uses only 5–20 MHz of spectrum. However, it is called wideband since the earlier version of CDMA was only 1.25 MHz in bandwidth. Thus, the words ultrawide, wide, and narrow are quite relative and dependent on the context.

the two technologies are currently at the same average power consumption (given that WiFi is quite mature and UWB is still in its infancy), the fact that UWB can deliver 5–10 times higher throughput than WiFi translates to that much more power saving in terms of joules/bit or watts/bps (watts per bit per second).

For instance, let us assume that both WiMedia UWB and WiFi can run at 500 mW of average active power. Then, transferring one gigabyte of data over UWB at, say, 100 Mbps of total throughput would take 1.33 min of time and 40 J of energy. Also, we can calculate that UWB takes 500 mW/100 Mbps = 5 nJ/b (nanojoules per bit transferred) or 5 mW/Mbps. For WiFi, the same gigabyte of data at 20 Mbps would take 6.67 min of time and 200 J of energy. In terms of energy per bit, WiFi burns 500 mW/20 Mbps = 25 nJ/b or 25 mW/Mbps, a fivefold increase over UWB.

The same comparison can be made with Bluetooth. Even though Bluetooth runs at an order of magnitude lower active power (~50 mW), its throughput is two orders of magnitude lower (1 Mbps) than UWB. Taking the same example of transferring one gigabyte of data, Bluetooth would take 2.2 hours and 400 J of energy. Bluetooth has a 50 nJ/b or 50 mW/Mbps power consumption rate.

Figure 1.1 illustrates the active power consumption of WiMedia UWB relative to Bluetooth and WiFi based on the above assumptions. One needs to keep a few points in mind relative to these assumptions, however. First, equating the average active power of UWB and WiFi is a bit unfair. WiFi technology and its integrated circuits (ICs) have had nearly a decade to evolve and improve. By comparison, UWB chips are in their first production release. Over time, it is expected that much lower powered UWB designs will evolve. Second, UWB throughput of 100 Mbps

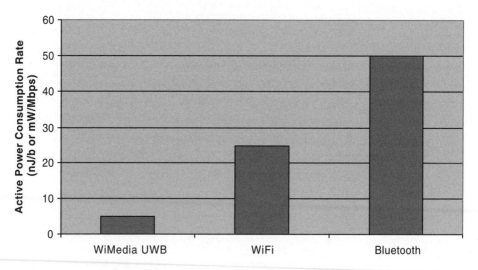

Figure 1.1   Active power consumption rate comparison

is by no means the best that can be expected. In fact, theoretically, the MAC sublayer should be able to deliver at the rate of over 350 Mbps at short ranges (3 m). Perhaps this is too high to expect from real implementations, but reaching at least a 200 Mbps throughput is well within reason. Thus, the power consumption gap between UWB and WiFi/Bluetooth will only increase over time, as UWB implementations mature.

Of course, one may argue that, with the advent of the next generation of WiFi (IEEE 802.11n), the data rate difference between UWB and WiFi will be minimized and, hence, the power consumption calculations above will not yield much advantage for UWB. However, the flaw in this argument is in the assumption that the next generation of WiFi will be consuming the same average active power as the old generation. 802.11n promises to deliver 100 Mbps of throughput for up to 100 m in range in only 20–40 MHz of bandwidth. To do this, this standard relies on

1. high transmit power;
2. high-order modulation schemes; and
3. highly complex digital signal processing (DSP) involving multiple-input multiple-output (MIMO) technology.

All these add up to higher size and cost (due to higher complexity) and higher power consumption.

Thus, owing to the extremely wide bandwidth of the UWB signals, they can achieve the high data rates with much less complexity and much higher power efficiency than any of the narrower band technologies could.

1.2.2 Little Interference to and from Narrowband Systems

Keeping the power levels lower than the noise floor of the other technologies would then allow UWB signals to coexist in the same frequency bands as the other technologies without generating much interference. Figure 1.2 illustrates the

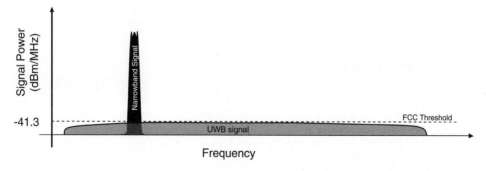

Figure 1.2   Relative power and bandwidth of UWB signals

Table 1.2 In-band emission limits of several narrowband technologies compared with UWB

Technology	In-band emission limit	
WLAN	5–50 mW/MHz	
WCDMA handset	63 mW/MHz	
Bluetooth 2.0	Class 3:	12 μW/MHz
	Class 2:	30 μW/MHz
	Class 1:	1.2 μW/MHz
UWB (USA)	74 nW/MHz	

relative power and bandwidth of UWB signals compared with more the conventional ones. It also shows the FCC emission limit set for UWB transmissions, -41.3 dBm/MHz. This is the same level as in FCC Part 15 rules, the unintentional power emission limits for all radiators. That is, UWB signals are as low in power levels as any other unintentional radiators (computers, hair dryers, etc.)

For comparison purposes, Table 1.2 gives the maximum in-band transmit power for several narrowband technologies in units of milliwatts per megahertz. It also includes the same for the UWB band in the United States. This puts in perspective the relative difference in power density between UWB and other technologies. Notice that UWB power density is in the nanowatts per megahertz range, whereas the others are running in the several microwatts per hertz (Bluetooth) to tens of milliwatts per megahertz (WLAN and WCDMA) range.

1.2.3 Unmet WPAN Demand

Figure 1.3 depicts the range and speed of high-speed UWB applications relative to those of WLAN and Bluetooth. This figure illustrates that high data-rate UWB solutions are needed to fill a gap that was not previously filled by any other technology. Other than Bluetooth, no other standard technology offered a WPAN solution. Since the current Bluetooth (version 2.1) is limited in its data rate, in essence there was no high-speed WPAN standard prior to UWB.

Short-range, high-speed wireless connectivity is an emerging application in all commercial sectors: PC, CE, and mobile communications. The convenience of wireless is desired (sometimes even required) by consumers, be it for their digital home entertainment system, personal computer and associated peripherals, or portable devices such as cell phones and digital still cameras. And, the multimedia content to be wirelessly transferred amongst these devices is richer than ever before, requiring higher connection speeds and quality of service (QoS) than afforded by available standard technologies. Clearly, the trend is towards higher throughput and lower latency requirements in the future.

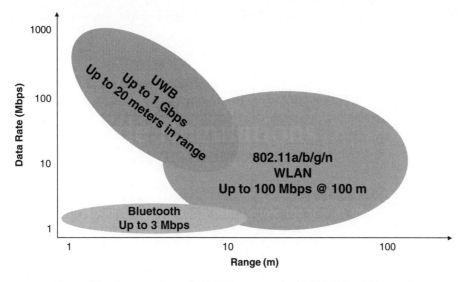

Figure 1.3 Range and speed of UWB compared with WLAN and Bluetooth

Hence, a wireless technology that can keep up with those requirements and concurrently keep the price, size, and power consumption down is highly desirable. UWB technology offers the potential to develop solutions that can meet these demanding requirements of next-generation WPAN applications.

1.2.4 High-precision Ranging

Owing to the high bandwidth of the UWB signals, their time-domain resolution is much higher than most other technologies. This allows the use of high-precision ranging mechanisms using UWB signals. The precision of such ranging systems can be on the order of centimeters. In comparison, the narrower band systems can at best offer precision on the order of meters.

1.3 UWB Market

When it comes to high data-rate connectivity and networking in the CE and PC markets, there are at least two major shortcomings that have been plaguing the consumer market. The first is the clutter of cables that has been commonplace behind most desktop computers and home entertainment equipment. The second is the lack of a reliable, easy-to-use, high-speed wireless networking capability.

1.3.1 Clutter of Cables

In this age of connectivity, there is hardly any consumer or computing device that does not need to be connected to another for synchronization, back-up, streaming, alerting, security, or other purposes. This connectivity has usually been in the form of cables, although recently the shift towards wireless has begun. The cables are of all sorts, and their type usually depends on the application. In the PC market, the USB cable is quite popular, especially for high data-rate applications. In the multimedia world (e.g. home entertainment), FireWire (IEEE 1394) may be more prevalent. Of course, many other cable types with differing performance and ease of use are showing up in the market everyday. These days, High-Definition Multimedia Interface (HDMI) cables are taking the center stage among all others in the home entertainment arena.

The cables are certainly very useful in their functionality; however, they have their down sides as well:

1. they are predominantly point to point;
2. they tend to be unsightly, especially when there is a large clutter of them;
3. they provide no flexibility (fixed length);
4. their in-wall installation is costly/troublesome.

Other inconveniences come about when considering portable devices. The cables used with these devices must always be carried around when connectivity is required. And, the fact that there is a myriad of different cable and port types makes the idea of matching a cable to a device daunting at times. Even within a cable family, say USB, there are a number of different physical interfaces that could make it incompatible. As a result, a wireless connection that can provide the benefits of the cable (high data rate, high security, ease of use) can be quite appealing to the average consumer.

1.3.2 Networking Shortcomings

The value of networking capability is also increasing these days. The desire to network different devices within a home, office, or even on our bodies is increasing steadily as different gadgets and applications find their way into our lives. Such networking is quite impractical over cables, especially if one considers portable devices. Consequently, the best networking option for such devices is wireless.

The existing wireless networking choices are: WWAN (e.g. cellular technologies), WLAN (WiFi), WPAN (Bluetooth), and certain proprietary technologies. The WWAN connections are usually costly for most consumers and the data rates are usually low. The Bluetooth option is free of service fees, but is also too low in data rate for use in multimedia and other high data-rate applications. The proprietary solutions rarely provide a networking capability, and when they do they are not interoperable among those from different vendors. Hence, so far, the most common and useful choice of wireless networking solution is WiFi (802.11 a/b/g), which can be free of service fees and comes with moderate data rates.

However, WiFi has its own shortcomings that make it unsuitable for certain high-speed WPAN applications. For example, IEEE 802.11a/g throughput is currently limited to less than 25 Mbps. This may be sufficient for some applications, but not enough for multimedia or other high-speed applications. More importantly, WiFi commonly requires an Access Point to be present[5] in the network. All communication among devices must be coordinated through the Access Point. This makes for a reduction in overall throughput of the network, as well as an increase in the latency of connections between devices. Also, the addition of the Access Point increases the overall cost of setting up the network. Most importantly, IEEE802.11 is a power-consuming platform that is not suitable for power-sensitive (portable) devices. As such, the use of WLAN in handheld devices (cell phones, cameras, etc.) is not user friendly due to the resultant short battery life.

Other shortcomings of WiFi for WPAN applications include:

- difficulty of setting up an ad hoc network on the fly;
- complicated network setup (the Access Point setup adds to the overall complexity);
- reduction of economy of scale in the market (two types of system development are required: Access Point and Device).
- lack of QoS in most WiFi systems (unless they implement IEEE 802.11e specification), reducing the usefulness of the network for streaming and other time-sensitive applications.

1.3.3 UWB Fills a Gap

As shown in Figure 1.3, the high-speed UWB technologies are needed to fill a large gap in the WPAN arena. With the advent of more and more speed-hungry applications (audio/video streaming, movie/music/game downloads, etc.) that run on all sorts of consumer electronics, computing peripherals, and mobile devices, the need for a high-capacity technology that can also be low power and low cost is very real. UWB has the potential to fill this gap very well.

Among the different high-speed UWB technologies, ECMA-368 [1] (from WiMedia Alliance, which we will also refer to as WiMedia UWB throughout this book) is clearly the winner, not just because of its technological merits, but more due to the political alliances formed in the industry around it. With industry giants such as Intel, Nokia, NEC, Samsung, NXP/Philips, STMicroelectronics,

---

[5] Ad hoc connections in WiFi are defined but rarely used.

etc., behind this technology, it has become the de facto standard of high-speed UWB.

By the same token, the alliance around WiMedia UWB creates the ecosystem of the variety of companies necessary for any new technology to be rapidly adopted in the market.[6] Since this is an unlicensed technology, it requires the cooperation and coexistence of many interoperable devices in the market for it to flourish. For instance, for the PC peripherals to contain UWB modules, their manufacturers need to see that the desktop computers and laptops are beginning to be equipped with UWB technology as well. This is currently taking place in the form of Wireless USB (W-USB), which is based on the adaptation of the familiar USB onto the WiMedia UWB platform.

The competing high-speed UWB technology came about through the efforts of Motorola/Freescale, leading to the formation of the UWB Forum SIG. However, ever since Freescale abandoned its interest and support for UWB altogether, the UWB Forum has lost its momentum and appeal, even though one or two companies are still pursuing the technology.

As such, WiMedia UWB is easily going to dominate the high-speed UWB market. The question is when and in what market segments. The timing has always been very difficult to predict in the fast-moving CE, PC, portable electronics (PE) market segments. And, as these segments are rapidly converging together, as shown in Figure 1.4, the distinctions among them are blurring. For example, a Smartphone these days can be considered a mobile communication device, but it is also a CE device for its multimedia capabilities, as well as a PC peripheral due to its PC synchronization and communication capabilities (not to mention its computing platform). Similarly, a personal digital assistant (PDA), a Bluetooth headset, or a digital still camera can no longer be considered as only a CE, a PC, or a PE device.

This convergence trend also confuses the choice of interconnect standards. In the PC market segment, the USB has been quite prominent as of late. However, in the CE segment, the FireWire (IEEE 1394) and HDMI standards have been more successful. In the PE segment, Bluetooth has had great success as a wireless connectivity means (for low data-rate applications), but USB has also found itself a foothold there, especially for downloads/uploads and synchronization. Of course, connection to the Internet is desired/required in most electronic devices as well, and the most popular method of connection to the Internet has been through Ethernet.

As we go towards a wireless cable-replacement technology, its ability to service the needs of two or more such legacy interconnect technologies comes in quite handy. In fact, that has been one of the main selling points of UWB: it can be

---

[6] For a list of WiMedia Alliance member companies that form this ecosystem, see Appendix B.

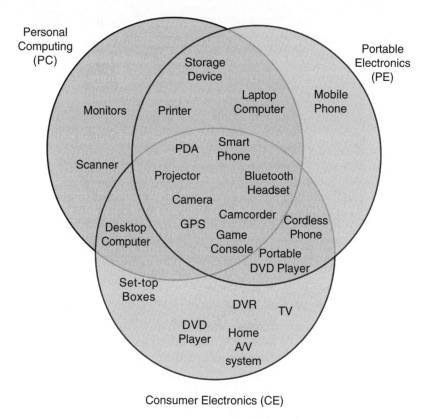

Figure 1.4   Convergence of PC, CE, and PE market segments

used as a platform for W-USB as well as wireless 1394, not to mention high-speed Bluetooth and wireless Ethernet bridges.

The WiMedia UWB platform has long been designed and promoted as the basis for the convergence of many upper layer stacks and applications (W-USB, High Rate Bluetooth, wireless Internet Protocol (IP), etc.) as shown in Figure 1.5. This figure is commonly referenced by the WiMedia Alliance and shows the intent of WiMedia UWB to act as the converging platform for multiple short-range technologies. This is accomplished by way of what WiMedia Alliance calls a PAL for each of the existing clients of the platform.[7]

The convergence of all the clients at the MAC sublayer is very important, since it provides for a guaranteed QoS over the common medium (wireless channel). This is true even though each of the MAC clients is unable to communicate with the other peer clients.

---

[7] Although the idea conveyed in Figure 1.5 is simple to understand, the implementation is not quite the same as drawn in the figure. See Chapter 5 for more details.

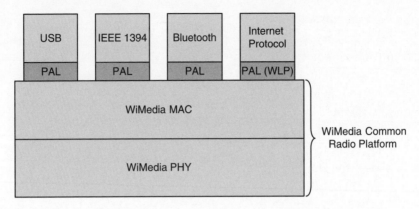

Figure 1.5 WiMedia platform allows for the convergence of otherwise unrelated connectivity clients

1.3.4 W-USB

One of the first use cases of WiMedia UWB technology has been considered to be the replacement of the USB cable, i.e. W-USB.[8] The USB Implementers' Forum (USB-IF, the SIG that developed the much successful USB standard), also heavily supported by Intel, was the first to adopt WiMedia as its platform (PHY and MAC) of choice for developing a wireless version of its USB technology, now called CW-USB. This has been a phenomenal boost for WiMedia UWB, opening the enormous USB market for adoption of WiMedia's technology. Billions of USB products are sold every year, and even if a fraction of them convert to CW-USB it will go a long way towards promoting the WiMedia UWB in the market place. The familiarity of the average consumer with USB technology and its reliable and user-friendly operation are considered to be the driving factors to get the consumers to upgrade to CW-USB. If the analysts' predictions are correct, then, in a few years, W-USB will replace USB in most computers, laptops, and a variety of CE devices.

With Intel heavily promoting CW-USB, analysts have expected the market acceptance to be rapid. This is true, especially because the CW-USB solution can theoretically provide up to 127 connections at the same time, as opposed to one for each wired USB port. Initially, the products will show up in the form of dongles (USB adaptors) and hubs. Of course, dongles are not expected to have the same performance as 'native' Hosts and Devices.[9] However, the simplicity of plugging

---

[8] The W-USB specifications are published by the USB-IF. However, there are other companies who have initiated the development of a wireless version of USB. In fact, Freescale was first to introduce a W-USB. However, there are major architectural differences between W-USB from USB-IF and those of the others. And, in order to differentiate its flavor of W-USB, the USB-IF started calling it CW-USB.

[9] Host and Device are terms used in USB and CW-USB protocols to refer to the master and slave of the bus respectively.

a dongle into an existing USB port in a computer or a device to turn it into a 'wireless' USB is very tempting and easily appreciated by the average consumer.

The CW-USB protocol is similar to USB, in that it follows the same master–slave networking approach, in which the Host dictates to the Devices what and when to send or receive. This type of approach typically allows for simpler Devices than Hosts.

Similar to the USB On-The-Go (OTG) specification, where a USB Device is given the capability to act as a simple (limited) USB Host as well, the CW-USB protocol allows for Dual-Role Devices (DRDs). A DRD can act as a Device, a Host, or both. DRDs should make it easier for CW-USB to infiltrate non-PC markets, especially the PE market.

Note that there are certain shortcomings to CW-USB. For example, the protocol is still very much PC-centric. The master–slave topology of CW-USB is another shortcoming, in that it reduces the overall throughput of the network by requiring one slave to talk to another one through the master (similar to WiFi), and it especially increases the power consumption of the Hosts (which must control all other Devices).

Despite these shortcomings, however, CW-USB is expected to have a tremendous market acceptance owing mostly to the reputation and market share of its wired parent. The market size estimates from various industry analysts range anywhere from 400 million to 500 million units by 2013. Of course, such long-range predictions come with a high degree of statistical variance; nevertheless, they show that the expectations are high for this market.

1.3.5 High Rate Bluetooth

Another familiar technology that has adopted WiMedia in its roadmap is Bluetooth. In 2006, the Bluetooth SIG announced that it will upgrade its popular technology to a high-speed one (termed Bluetooth 3.0, also known as High Rate Bluetooth) based on WiMedia UWB technology. This news generated quite a buzz in the industry, since Bluetooth, like USB, has amassed quite a following, mostly in the mobile phone and other PE devices. If successful, this name recognition also helps WiMedia to quickly find a footing in PE, especially in the mobile phone market.

By the same token, the Bluetooth protocol stack can also benefit from the addition of a high data-rate platform to the otherwise low-rate technology. In addition to the data rate increase, the benefits include:

- considerable power efficiency in high-rate applications;
- potential for new Bluetooth profiles for high-rate applications;

- potential for enhanced and more sophisticated networking features;
- range measurement capabilities.

Bluetooth 3.0 specification requires that each Bluetooth device be backwards compatible with Bluetooth 2.1 (which in turn is backwards compatible with earlier Bluetooth versions). This means that each Bluetooth 3.0 device will contain at least two MAC and PHY platforms: one based on the old Bluetooth MAC and PHY and one based on an alternate, high-rate MAC and PHY.

Bluetooth 3.0 is architected to be able to interface to any approved MAC and PHY platform for high-rate operation. Bluetooth terminology refers to this high-rate platform as the Alternate MAC–PHY (AMP). This also allows for IEEE 802.11 to be a contender for the AMP in High Rate Bluetooth [3]. In fact, given the popularity and political momentum of WiFi, it has become the main contender for AMP for the next generation of Bluetooth. However, given the issues associated with IEEE 802.11 AMP (high power consumption, throughput lower than UWB, limited QoS, etc.), it is unlikely that it will be able to displace WiMedia UWB in the future Bluetooth market altogether. Interference issues with certain fourth-generation (4G) mobile communication and WiMAX licensed bands in the 2.3 and 2.5 GHz bands (adjacent to the 2.4 GHz band of IEEE 802.11) can also be of concern and require further study [4].

1.3.6 IP over UWB

The WiMedia SIG has also been working on an IP adaptation layer (or, more accurately, a logical link control sublayer) called WLP that would essentially allow IP-based technologies (e.g. DLNA) to use WiMedia MAC. This would then provide a path to IP-based applications that are currently prolific in the market. This is also considered a very important roadmap for WiMedia. It is the path to IP and layers above it (TCP, UDP, ICMP, FTP, HTTP, DHCP, SOAP, UPnP, DLNA, etc.). As of the time of writing, the specification for WLP has not been published in Ecma International. However, it is expected very soon.

The IP-enabled consumer electronic market is growing very rapidly, thanks to the ubiquity of IP. IP connectivity has been mostly dominated by Ethernet. About half of all IP-enabled devices sport an Ethernet port, whether or not they also have other types (e.g. wireless) of IP port. However, the other contenders for IP connectivity are making rapid progress towards taking a larger portion of the market share.

Among the IP-enabled CE devices, gaming consoles hold first place. By some accounts, over 5 million users played videogames over the Internet in 2007. These numbers are expected to grow rapidly in the coming years.

Among wireless IP connectivity solutions, WiFi (IEEE 802.11a/b/g) holds the lion's share. However, the throughput and QoS limitations of WiFi leave much more to be desired. Although IEEE 802.11n is expected to solve some of those issues, the high power consumption and cost will remain a major shortcoming when it comes to portable (battery-operated) devices.

IP over UWB could be a very useful technology for short-range applications. It will offer the bandwidth required by the ever-increasing multimedia applications that require IP connection (e.g. video gaming, video over IP, wireless high-definition displays), while keeping the power efficiency very high (i.e. energy consumption per bit very low). Moreover, the short range of IP over UWB could actually be an asset in certain circumstances. For instance, having a very high-speed IP connection over WiFi in a room for gaming purposes could be difficult when other high-speed IP connectivity applications are taking place within a home environment. The short range of UWB provides the spatial diversity necessary to allow each room in the house to have its own high-bandwidth connections without competing (interfering) with each other.

Since WLP is developed by the same people who have developed the MAC sublayer of WiMedia, it is very well matched to the capabilities of WiMedia MAC. This affords WLP a potential performance capability above that of W-USB and High Rate Bluetooth. Plus, it does not have the extra burden of carrying two MAC and PHY platforms, like High Rate Bluetooth. Thus, a WLP-based solution should be quite superior in performance and cost, and potentially in power. Nevertheless, the battle to gain market share for WLP will be an uphill one. There is no major support for WLP outside of WiMedia. Most attention has been paid to W-USB and Bluetooth, as customers of WiMedia. Hence, WLP has to prove itself worthy before it will be noticed.

1.3.7 What to Expect

In the early days of WiMedia, many use cases were drawn up with the hopeful and aggressive plan of getting them marketed fast. The ideas and applications are still valid; however, expectations have been tempered with the realities of getting a new technology out of the laboratories and into the marketplace.

It has been so for any new technology that tried to debut itself, be it Bluetooth in WPAN, IEEE 802.11 in WLAN, or CDMA in WWAN. Delays and temporary shortcomings in new technologies are part of the process of introducing a new technology to the market. However, as is always the case, critiques are abundant pointing to shortcomings in the early days. These critiques usually come with short historical memory, failing to remember similar obstacles that the now-prolific technologies had to overcome. Bluetooth and WiFi have been two relatively recent technologies that went through very similar challenges in their debut years, where initial high hopes were shattered with technical and market setbacks. Now, it seems as if we cannot fathom being without

WiFi. And, more than 1 billion Bluetooth devices were in service by end of 2006.

The current predictions are that by the year 2013 the number of CW-USB products in the market will rise to around 400 million. Whether this is an accurate prediction or not, it shows the optimism of the market towards this technology. Adding Bluetooth, WLP, and other potential clients of the WiMedia platform, it is easy to see that this technology is destined to have a respectable market share in the WPAN arena.

1.4 UWB Standardization

Let us go back in time and review the history of UWB standardization. This will allow a better appreciation of the current status of the UWB standards.

On 14 February 2002, the FCC of the United States approved a spectrum in the range 3.1–10.6 GHz to be available for the unlicensed use of UWB signals. This approval was pivotal in creating the momentum in the industry to proceed with the development of UWB standards. Soon after, other regulatory bodies started following suit. Of course, each country or region has its own flavor of the regulations dictating the UWB signal transmissions and spectrum. However, the worldwide acceptance of UWB technology as a viable and necessary means of personal area communications and networking was established. As such, the necessity for the development of standardized specifications for the operation of such a technology became even more apparent.

1.4.1 IEEE

The pioneering UWB companies attempted the standardization process within the IEEE organization. IEEE Standards has been one of the most respectable standards organizations, especially for unlicensed wireless communications. IEEE 802.11 is a very common term used to refer to the very successful WLAN technology. Even lay people are now very familiar with the term 802.11 and its associated SIG, i.e. WiFi. Hence, it was only natural to bring the UWB standardization effort to the IEEE Standards organization. Two task groups within the IEEE 802.15 Working Group for WPANs were initiated: Task Group 3a (802.15.3a) for high data-rate UWB (the focus of this book) and Task Group 4a (802.15.4a) for low data-rate UWB applications.[10]

---

[10] Examples of these applications are industrial inventory control, home sensing, control and media delivery, logistics, industrial process control and maintenance, safety/health monitoring, personnel security, low data-rate communication.

In December 2002, IEEE 802.15.3a was mandated by a Project Authorization Request (PAR) to provide a higher speed, UWB-based PHY for multimedia applications. The task group started with 23 PHY proposals and was quickly able to consolidate (down-select) them down to only two competitors: MB-OFDM or Direct Sequence CDMA (DS-CDMA). Further progress was not achieved, however, as the process was hampered by the dichotomy in the industry on the types of signal used in the two proposals. Each camp showed simulation and prototype results that claimed superior technological performance. The alliances had, by this time, turned into more of political fronts between the two technology choices. Neither group showed any willingness to compromise. For many months, the two camps provided sufficient member participation in the IEEE 802.15.3a Task Group meetings to prevent each other from gaining a super majority (70%) vote. Finally, on 19 January 2006, the task group members voted to withdraw the PAR, ending the prospect of an IEEE standard for the high-speed UWB technology.

The low data-rate UWB technology standardization in IEEE 802.15.4a did not face such a fate and went through to completion.

1.4.2 WiMedia versus UWB Forum

While battling each other in the standards meetings, and before the dismantling of the standardization effort in IEEE 802.15.3a, the proponents of the final two proposals had each created their own SIGs. One was called the MB-OFDM Alliance (MBOA), which, as the name suggests, supported the MB-OFDM proposal. This SIG was led by Intel, Texas Instruments, Nokia, STMicroelectronics, Hewlett-Packard, Philips (later NXP), Samsung Electronics, Sony, and some UWB-specialized start-up companies. Later, CSR (formerly Cambridge Silicon Radio) joined the group as well.[11] MBOA later merged into the WiMedia Alliance (WiMedia for short), which was originally created as a nonprofit organization to promote applications based on the MBOA's technology. Since the same companies were on the boards of both SIGs, it made sense to join these two groups into one.

The opposing SIG was called the UWB Forum and was mostly driven by Freescale (a spin-off of Motorola). This camp was supporting the DS-CDMA, a technology very near and dear to Motorola (with plenty of intellectual property to go with it) due to the fundamental similarities to the CDMA technology that Motorola had been developing for many years in the mobile communication industry.

---

[11] Texas Instruments is no longer a member of WiMedia, as of writing. Microsoft had also joined as a board member for a while, but has recently reduced its activities to the contributor level.

Even though the standardization process was effectively stalled within the IEEE Task Group, each of the two SIGs continued the development of their own detailed PHY and later MAC specifications in earnest. By the time the IEEE Task Group was dismantled, the two camps had almost completed their respective PHY specifications.

The industry was quite frustrated with the lack of a compromise solution to merge the two competing standards into one, reminiscent of the old VHS–Betamax industry deadlock. The WiMedia Alliance camp, having been discouraged about the prospects of an IEEE standard, decided to publish its PHY and MAC specifications in Ecma International – European association for standardizing information and communication systems (Ecma, for short). Once accepted by Ecma for publication, WiMedia's UWB technology finally achieved the credibility of a recognized standards body.

Meanwhile, the UWB Forum unexpectedly ran out of steam when its main leader, Freescale, gave up on UWB technology altogether. As a result, the UWB Forum practically vanished in terms of a viable competitor to the WiMedia Alliance. Unfortunately, this development happened after the disbanding of the IEEE 802.15.3a Task Group. By this time, WiMedia had already standardized its technology within Ecma, and it was too late to go back to IEEE. Nevertheless, WiMedia UWB is currently recognized as the de facto high-speed UWB standard,[12] and it no longer matters much which standards body is sponsoring it.

1.4.3 Ecma International

Founded in 1961, Ecma International used to be called ECMA (European Computer Manufacturers' Association) until 1994. Ecma International's charter aims to facilitate the standardization of technologies in information and communication technology as well as in CE. Ecma provides access free of charge or copyright to all its standard documents.[13]

WiMedia submitted its specifications to Ecma in 2005. In the same year (8 December 2005), Ecma approved the publication of the combined PHY layer and MAC sublayer specifications in a single document: ECMA-368 [1].

In addition, WiMedia's MAC–PHY Interface (MPI) specification has been published as ECMA-369 (WiMedia MAC–PHY Interface) [5], which proposes a possible interface between PHY and MAC. This MPI specification is not mandatory for the manufacturers to follow. However, it does offer a point of

---

[12] The DS-CDMA-based UWB is still being pursued by some companies in some applications, such as home entertainment.

[13] http://www.ecma-international.org.

interoperability between PHY and MAC IC developers. That is, it allows PHY ICs and MAC ICs to be independently developed by different companies without running the risk of interoperability problems between them when they are integrated into a single system. Owing to the fact that the trend for WiMedia UWB is in single IC solutions (to reduce cost), the MPI specification may not play a major role in the future. However, it has so far been well received and adopted by most UWB PHY developers.

Since their initial publication, ECMA-368 and ECMA-369 have undergone a revision. In December 2007, Ecma released their second editions. In the future, it is expected that other WiMedia-related specifications (e.g. WLP, WiMedia's logical link control sublayer) will be published within Ecma as well.

The ECMA-368 standard has also been published by the International Organization for Standardization (ISO) and the European Telecommunications Standards Institute (ETSI): ISO/IEC 26907 and ETSI IS 102 455. Moreover, for ECMA-369, there is an ISO counterpart: ISO/IEC 26908.

Throughout this book, we will use the terms 'ECMA-368,' 'WiMedia MAC and PHY,' and 'WiMedia UWB' interchangeably.

1.5 UWB Applications

This section is dedicated to the applications (or clients) of the WiMedia UWB platform. In this section, we shall consider the term 'application' to be from a user (market) perspective, and not necessarily from the strict sense of the Open Systems Interconnection (OSI) model (see Section 1.7).

UWB applications have generally run the gamut from the low-speed data collection and ranging scenarios to very high data-rate wireless HDMI. The range of applications also goes from the obvious to the highly imaginative.

WiMedia's general approach for targeting applications for its UWB technology has been to associate it with the successful technologies (clients) that are otherwise either wired or low throughput. Examples of such technologies that could use the wireless flexibility or the high throughput offered by the WiMedia UWB platform are:

- USB;
- IP and related upper layers (TCP, DLNA, UPnP, etc.);
- Bluetooth;
- FireWire (IEEE 1394).[14]

---

[14] The 1394 Trade Association, which maintains the IEEE 1394 standard, had originally agreed to the use of WiMedia UWB as its wireless platform, but has changed its mind since. Thus, the work on the corresponding PAL has been halted.

These are not the only applicable areas/applications, of course. However, these have been the focus areas for developing PALs for the WiMedia platform. In fact, these are reflected in WiMedia's commonly referenced platform layering diagram, as shown in Figure 1.5. Although this diagram is a bit misleading in terms of the clear interfaces that may exist between WiMedia's MAC sublayer and its PALs (see Chapter 5 for a treatment of PAL interfaces), it is a clear indication of where the general trend within WiMedia has been when it comes to enabling applications.

The most popular of such application ideas is W-USB. Enabling W-USB was, and still is, the first target of the WiMedia Alliance. The reason for this is that the market for W-USB seemed, and still seems, ripe. Billions of USB devices are sold every year, especially on personal computers and PC peripherals. They generally work quite predictably and without any installation requirements. They run at very high data rates (480 Mbps is the USB rate at its physical layer, but its maximum throughput at the driver level is at best 250 Mbps). The plethora of applications that have been enabled through the use of USB connections are not limited to personal computers and PC peripherals. Nowadays, USB ports can be found on mobile phones, PDAs, and digital cameras/camcorders, sometimes even doubling as the power source for the portable devices. Thus, the average consumer of a PC or PE device has been well conditioned to the use and experience of USB.

Reliance on the USB connection is now so prevalent that the cable clutter behind a desktop computer or a laptop docking station is, for the most part, due to all of the USB cables running to the peripherals. Therefore, the promise of 'cutting the cords,' so to speak, with the use of a W-USB technology is not only pleasing, but almost necessary. It is not uncommon for a consumer to run out of USB ports on their desktop computer or laptop, requiring the procurement of a USB hub. Limitations on the USB cable length has also had a restrictive effect on the average consumer. If the data rate and reliability can be kept the same as in USB, and the price low, then what is to keep the average consumer from making the transition from wired to wireless? It just seems elementary, at least to a layperson. (The short range of WiMedia UWB does not matter here, since USB cables are even shorter.)

Hence, the CW-USB was envisaged by WiMedia, USB-IF, and others in the industry as the lowest hanging marketing fruit. Even lower was the idea of using dongles (wire adaptors) to turn any USB port into a 'wireless' USB port. Of course, personal computers and PC peripherals were the first area of focus, but specifications for the non-PC use of W-USB have been in the works for some time, as well.

Figure 1.6 illustrates some of the basic W-USB applications that can be imagined. These include mainly peripherals connecting to the personal computer. In addition, the synchronization and downloading functions between a personal computer and certain portable devices, such as digital cameras/camcorders, mobile phones, PDAs, etc., would fall in this same category.

Figure 1.6 Basic W-USB applications

Figure 1.7 depicts some of the computer-less applications of W-USB. In these applications, there is no personal computer involved. A Device (DRD) may act as a Host to another device.

IP over UWB has also been an idea that was formed from the early days of WiMedia. In fact, WiMedia took it upon itself to create a PAL to IP, currently named WLP.[15] The notion was that, since WiMedia MAC creates such a high-efficiency, QoS-enabled, ad hoc, peer-to-peer network capability, it would be quite synergistic to tie it to the ever-so-popular and similarly ad hoc and peer-to-peer network of the Internet through the use of IP. There is no need for IP within a group of WiMedia devices communicating through the MAC sublayer; however, at some point the necessity for a connection to the Internet/Ethernet-based applications is likely. Having a mechanism to make this connection seamless is quite beneficial, both in terms of complexity and throughput. WLP uses IP packets to transmit over WiMedia MAC. As such, all IP-based protocol stacks and applications can easily be using WiMedia UWB as the transport mechanism.

Figure 1.8 illustrates examples of IP-over-UWB applications possible via WLP. Audio/video streaming over IP (e.g. voice phones) is an especially attractive application. Because there is no Host–Slave relationship, the WLP-based connectivity

---

[15] WLP used to be called WiNet.

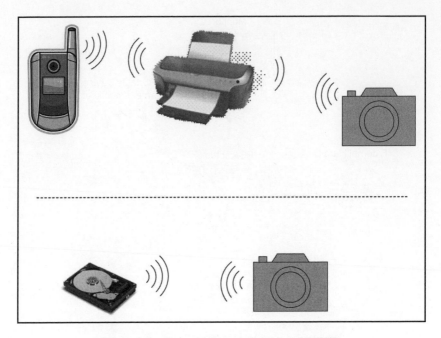

Figure 1.7   Computer-less applications of W-USB

among the devices is point to point, creating a more efficient mechanism for devices to stay in touch with each other.

Figures 1.9 and 1.10 give two more examples of WLP-based applications. Figure 1.9 depicts a remote gaming scenario in which WLP provides the transport among local devices as well as those remotely connected through the Internet.

In Figure 1.10, an ad hoc conference room or classroom is illustrated in which no network preplanning is performed, or in which no permanent network connections among the different devices are created. Conference participants can walk into the room and join the WLP-based network of devices with each other, as well as with the WLP Bridge that connects them to the Internet. In the figure, this bridge is also a WLAN Access Point that makes the connection to the Internet for the WPAN. However, other arrangements, such as connecting a WLP Bridge to the wired LAN in the office (Ethernet), or even the WiMAX WWAN are also feasible.

Compared with USB, IP, and IEEE 1394, Bluetooth is an exception in that it is already a wireless technology. Successful adoption of Bluetooth in the market place has been mainly due to mobile headphones and hands-free applications. The low-power nature of Bluetooth (for low data-rate applications) is quite appreciated in portable devices, especially mobile phones. The popularity of Bluetooth in the mobile industry has recently spilled into other industries as well. You can find Bluetooth in laptops, mice and keyboards, global positioning system (GPS) receivers, MP3 players and watches, and stereo headsets and sunglasses. Bluetooth

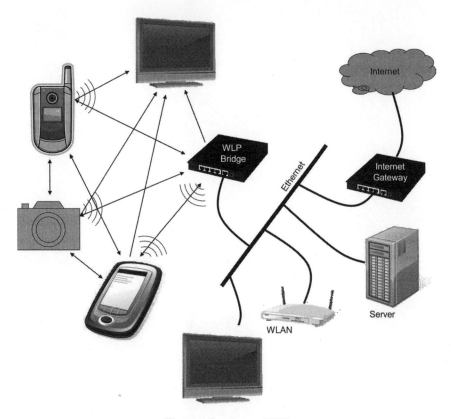

Figure 1.8   IP over UWB

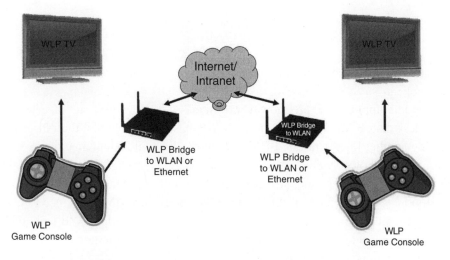

Figure 1.9   Example of IP-over-UWB application: remote gaming

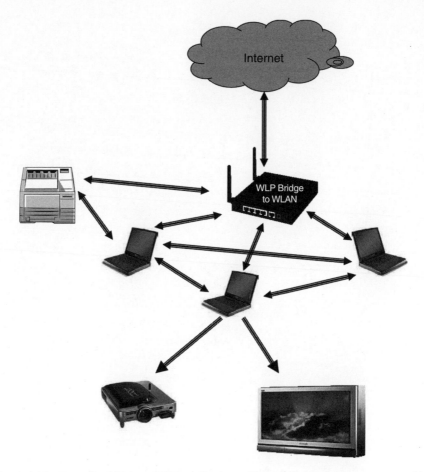

Figure 1.10   Example of IP-over-UWB application: ad hoc conference/class room networking

dongles have also been manufactured for a variety of data and voice communication applications.

Bluetooth applications are, however, limited by the low data rate of Bluetooth
PHY (less than 3 Mbps). The Bluetooth SIG has wisely decided that, instead of
creating another high data-rate PHY technology, it should adopt an existing ones.
The obvious choices were UWB and IEEE 802.11. Among UWB technologies,
WiMedia's was chosen since it has become the de facto standard in the industry.
Bluetooth's announcement of this choice was a boost for WiMedia, since it meant
that mobile phone applications can take advantage of WiMedia UWB technology
as their high-speed transport mechanism. This could open the door for WiMedia
UWB semiconductors in the mobile handsets.

Of course, the choice of UWB bodes well for the Bluetooth SIG as well. By
transitioning to a 480 Mbps (PHY) technology, Bluetooth can open the door to a

set of high-speed applications that it would not be able to service otherwise. The high power efficiency of the WiMedia platform is also consistent with the tradition (and expectation) of Bluetooth applications.

The Bluetooth SIG has defined an AMP roadmap whereby high-speed wireless technologies such as IEEE 802.11 and UWB can be added to the Bluetooth Core specification. That is, the AMP does not replace but augments the Bluetooth transport offering. In fact, the original Bluetooth radio technology (version 2.1) is mandatory in any future Bluetooth device for the purpose of backwards compatibility. It is used for discovery of peer devices and their capabilities, for authentication, and for low data-rate communication. The AMP link will be used as needed for high-speed communication; otherwise, it will be turned off or put in idle mode.

The Bluetooth SIG has designed the new AMP-enabled architecture (High Rate Bluetooth) by making sure all upper layer stacks can remain unchanged (as much as possible). Also, the SIG has left the door open for any future alternate technology to be used as its transport. That is why IEEE 802.11 (WLAN) technology is also being considered as an AMP for the next version of Bluetooth. Even though WiFi has a higher power consumption and lower data throughput than UWB, IEEE 802.11 enjoys a much higher market acceptance and wider coverage area. By allowing multiple AMPs to compete for market acceptance in a Bluetooth device, the SIG is allowing the market to decide the best choice for the high-speed link.

Applications of High Rate Bluetooth over UWB that have been contemplated so far revolve around the use of a Smartphone or PDA in high-rate applications. These include, but are not limited to:

- downloading/uploading of videos/movies/songs to/from Smartphone/PDA more than 100 times faster than before;
- streaming high-definition audio/video/presentations from one device to another;
- playing games on the device while connected to a large-screen TV or monitor;
- streaming the video output of any Bluetooth device to a larger screen;
- streaming the audio output of any Bluetooth-enabled portable device to a high-fidelity sound system;
- high-speed connection to the Internet and related applications, such as video streaming;
- audio/video streaming and downloading in to an in-car entertainment system;
- back-up and synchronization of data from Smartphone/PDA to hard disks, personal computers, etc.

With or without W-USB, WLP, or High Rate Bluetooth, the WiMedia platform can be used to enable a variety of use cases. One of the more interesting ones currently being pursued by many companies is that of the home entertainment network. Similar to the PC environment, UWB can offer cable replacement (and more) for an audio/visual entertainment system in the living room. The high throughput of WiMedia MAC is quite useful in enabling high-definition video transmissions among set-top boxes, TVs, DVD players, and other such devices, with moderate (lossless) compression. High fidelity 7.1 surround sound can also be carried over the same network to wireless speakers around the room. Moreover, portable devices (such as Smartphones, laptops, PDAs, game consoles) can engage with the same network in both loading and streaming of data from one to another as needed.

The short range of UWB can be a benefit in such situations where the high-bandwidth applications are localized. The reason is that spatial diversity allows another UWB system in another room to operate with the same full capacity without having to share the overall bandwidth. This is not the case when, say, WLAN is used in a home entertainment application. Owing to the wider range of WLAN, its application bandwidth capacity will have to be shared among all users in its range. This means that if much of the bandwidth is used up for the high-definition streaming in the living room, then not much will be left for gaming and Internet access in the other rooms in the house.

A better approach to the house or office building coverage is one where WLAN and UWB cooperate to cover both the range and bandwidth requirements. In effect, WLAN and or LAN would act as the overlay/backbone technology connecting pockets of UWB networks to each other and to the Internet. Imagine the home environment of Figure 1.11 where a LAN and/or WLAN covers the entire house. Within each room a group of UWB devices can communicate with each other without taking up the scarce capacity of the LAN. Hence, the LAN will still be able to service all rooms with a reasonable application bandwidth.

Related to this topic there are other, as yet not standardized and less-talked-about areas of WiMedia UWB application, such as wireless HDMI and UWB-over-Coax. Wireless HDMI tries to make the now-popular HDMI connections wireless. Of course, uncompressed high-definition video does not fit the bandwidth offered by the current version of WiMedia UWB.[16] However, several companies have enabled such technologies using lossless[17] (e.g. JPEG2000) compression/decompression techniques.

UWB-over-Coax is a relatively novel approach in which some of the commercial video carriers are interested. Since most of the houses in the US come wired

---

[16] A revision with higher data rates is expected from the WiMedia Alliance soon.

[17] The use of lossless compression/decompression is necessitated by the encrypted content being transferred by the application. The Digital Rights Management (DRM) and copy protection requirements of most digital movies, songs, and games dictate the use of encryption.

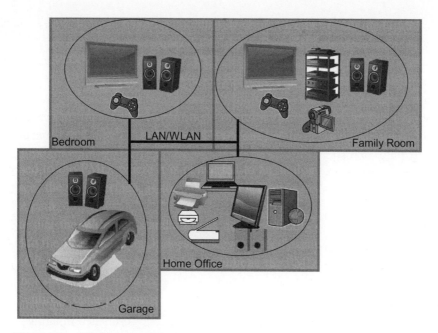

Figure 1.11   WPAN and LAN/WLAN cooperation in a home entertainment application

with coax cabling, it would be beneficial to take advantage of that wiring for send-ing UWB signals. The advantage of this approach is multiple-fold:

- Non-cable service providers will be able to offer video service throughout the home without extra wiring costs.
- With little modifications, the same technology can be used to transfer signals over coax as it would be to carry over the air. This allows for
  – economy of scale in the lowering of the price of the UWB devices manufac-tured for both air and coax purposes;
  – reuse of the MAC (and upper) layers in going from coax to air or vice versa, avoiding another layer of protocol adaptation that could reduce the overall throughput.

With UWB-over-Coax, the range shortcomings of UWB can be mitigated in certain scenarios.

In areas of the world where coax cables are not commonly found in buildings, the use of UWB over power lines may be a viable option.

So far, we have focused on applications related to home and office. In all of the above, the consumer is interested in replacing wires among devices. The most data-intensive applications are in the area of high-definition video streaming to a TV/monitor. The source could be a set-top box, a DVD player, a game console, a camcorder, a mobile phone, etc. The cable replacement in these cases is mostly due to the fact that the user would like to improve the convenience or aesthetics of the environment. However, the networking capability is also a major factor in the decision-making.

The same set of video applications can be envisaged in the automotive applications, where a network of highly interconnected devices can transfer data and stream voice/audio/video to each other. In-vehicle entertainment systems are becoming quite sophisticated, with their own set of challenges. UWB can easily meet the range requirements in these types of environment. Plus, it can offer the required bandwidth and QoS for all in-vehicle applications.

Another area of applications for WiMedia UWB is in medical industry. The wireless medical and healthcare applications have been growing rapidly. With UWB, such advancements can be further speeded up.

Some examples of medical applications are:

- streaming of video/audio among medical devices;
- wireless computed axial tomography scan, magnetic resonance imaging, ultrasound, X-ray, endoscope, etc.;
- wireless data and image transfers.

These may be suitable for a doctor's office just as well as a hospital environment. High reliability and QoS are key here compared with cost or power consumption in the consumer market. The data rates can be very high and the latencies tolerated are usually very low. As long as the range of operation is short, then a UWB solution should be able to accommodate such requirements.

Hospital entertainment systems are becoming popular as well. Services such as telephone, standard and premium TV and radio programs, games, Internet and email are being offered over wired networks. Similar to the home entertainment application of Figure 1.11, UWB can easily make the offering within each hospital room completely wireless to reduce the clutter of wires and the associated cost and inconveniences.

Surveillance and monitoring are also prime targets for UWB applications. These include:

- industrial digital wireless surveillance cameras and monitoring systems;
- CCTV in home or store;
- baby/patient/elderly monitoring.

These applications can benefit from the WLP adaptation to the IP network, allowing the monitoring to be performed and/or stored remotely over the Internet.

1.6 Summary Features

At this point, it would be useful to have a general view of the features and capabilities of the WiMedia UWB PHY and MAC. These highlights will help explain what sets WiMedia UWB apart from alternatives.

Our reference will be ECMA-368 [1], which specifies the PHY layer and the MAC sublayer for short-range, high-speed wireless communication and networking. It is based on the spectrum allocated by the FCC in the United States in the range of 3.1–10.6 GHz.

The spectrum is divided into 14 frequency bands of 528 MHz each. Six band groups are defined based on various groupings of such bands together, as shown in Figure 1.12. The first four band groups consist of three bands each. The fifth band group contains only two bands. The sixth band group was later defined as three-band group overlaid on top of Band Groups 3 and 4 in such a way to better match the requirements of most (if not all) regulatory bodies around the world.

To alleviate regulatory concerns over interference to licensed operations in the UWB bands (such as 4G mobile telephony and WiMAX), tone-nulling and interference avoidance mechanisms have been incorporated into the PHY and MAC specifications.

The PHY portion of the specification provides for a short range communication at variable data rates ranging from 53.3 to 480 Mbps.[18] Time- and frequency-domain spreading and puncturing are used to vary the data rate. The modulation is based on a MB-OFDM scheme. The OFDM operation uses 100 data carriers and 10 guard carriers in each band. The transmission is then hopped sequentially and periodically over the different bands within a particular band group, increasing the effective bandwidth of operation to 1.5 GHz for Band Groups 1, 2, 3, 4, and 6, and to 1 GHz for Band Group 5. Different hopping patterns are defined that are the basis for channelizing the band group. (Several time–frequency codes (TFCs) are available to allow for multiple logical channels within a band group.) For the purpose of synchronization and tracking, each PHY packet contains a preamble as well as 12 pilot subcarriers. To combat various wireless channel impairments, frequency-domain and time-domain spreading and forward error correction (FEC) – convolutional encoding, puncturing, and interleaving – are provided. The packet header is protected with a 16-parity-bit cyclic redundancy code (CRC) and the payload with a 32-parity-bit one.

WiMedia PHY mandates the implementation of data rates 53.3, 106.7, and 200 Mbps, as well as all logical channels defined by TFCs. At least one of the band groups must be implemented as well. (Band Group 1 used to be mandatory in the first version of the standard, but no longer.)

---

[18] The next version of this standard is expected to easily double the maximum data rate.

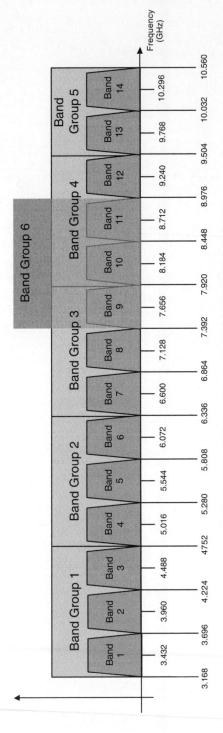

Figure 1.12 Bands and band groups defined by WiMedia PHY

The MAC sublayer is quite versatile in mobility, allowing individual and groups of devices to join or leave a network. All devices are peers to each other; as such, the overall MAC control of the network is equally distributed among all devices. (No master–slave relationship is defined.) Each device is expected to send a beacon frame every superframe (a time slot about 65 ms long) to allow for coordination of devices in the network. Most MAC control mechanisms are accomplished through the use of beacon messaging (Beaconing).

To deliver the QoS required by all types of application traffic (audio, video, bulk transfer, etc.), the MAC sublayer provides medium access through a sophisticated Distributed Reservation Protocol (DRP), a form of Time Division Multiple Access (TDMA). This is in addition to the more traditional Carrier Sense Multiple Access (CSMA) type of medium access, termed Prioritized Contention Access (PCA) in this standard.

Each superframe is made up of 256 time slots of 256 µs each. The Beaconing protocol allows each device to reserve a set of slots for unicast, multicast, or broadcast purposes. The reservation process prevents competition among devices for medium access and, therefore, reduces the overall MAC overhead, increasing MAC-level data throughput. The reservation also allows for guaranteed service to applications that require it, such as streaming video/audio.

The use of Beacons and DRP has another side benefit. It allows devices to send and receive all their coordinating messages (Beacons) within a relatively short period of time at the beginning of each superframe (Beacon Period), and conserve energy during the rest of the superframe as much as possible. In contrast, WiFi's CSMA-based MAC protocol requires a device to examine the channel usage constantly before sending its own data, causing unnecessary power consumption when the channel is relatively busy.

In addition, the MAC protocol allows for device hibernation, which lets devices skip all Beacon activity (along with all other transmissions and receptions) for an extended period of time (many superframes). This provides for further reduction in the standby power consumption of power-sensitive devices.

Although an optional feature, payload security using an AES-128 (Advanced Encryption Standard[19] using 128-bit blocks) encryption/decryption engine with CCM (Counter with CBC-MAC[20]), as well as replay attack protection is provided in the MAC protocol. Moreover, a four-way handshaking mechanism allows for establishment of session keys based on shared private keys.[21]

---

[19] A cryptographic algorithm approved by Federal Information Processing Standard (FIPS).

[20] CBC-MAC stands for Cipher Block Chaining Message Authentication Code, which is one of the methods for checking message integrity.

[21] The mechanism of creating a shared private key is beyond the scope of the MAC sublayer and involves the application layer (and the user).

The WiMedia MAC is designed to be a convergence platform to allow multiple, independent, MAC clients to coexist in the medium and provide the necessary QoS without causing interference to each other. For now, CW-USB, High Rate Bluetooth, and IP over UWB (WLP) are planned to use WiMedia MAC as their platform.

Hence, the WiMedia MAC and PHY are designed to provide very high throughput, low power, true ad hoc, peer-to-peer WPAN capability, with guaranteed QoS and a high focus on coexistence and mobility. The short range (about 10 m) of the PHY has the benefit of allowing spatial reuse of the UWB frequencies in short distances.

1.7 Terminology

WiMedia specifications currently include PHY, MAC sublayer, MPI, WLP, PHY certification, and Platform certification. These specifications are privately held by the WiMedia Alliance membership until they are published in Ecma International or other standard bodies.

CW-USB is a standard developed and published by the USB-IF. This protocol builds on the WiMedia PHY and MAC as a platform.

In this book, we will focus our attention on the specifications included in ECMA-368 [1] standard – WiMedia PHY layer and MAC sublayer – as well as the CW-USB specification [6]. The focus of this section is to give the necessary background and terminology to understand the language of these standards. We will also describe the terminology used consistently throughout this book.

WiMedia specifications frequently use the nomenclature of ISO/OSI-IEEE 802 Basic Reference Model [7] (which is also known simply as the OSI model) for the hierarchical communication architecture. Figure 1.13 depicts this model. Without some basic understanding of this model and the terminology used in it, it is difficult to follow some parts of the ECMA-368 specification.

The focus of this chapter is on the first two layers: PHY and the Data Link Layer (DLL). The latter is further split into the MAC sublayer and Link Layer Control (LLC) sublayer. The ECMA-368 specification limits itself strictly to PHY and the MAC sublayer.

Further expanding the first two layers, Figure 1.14 illustrates the PHY, MAC, and Device Management Entities (DMEs), the different Service Access Points (SAPs), and their relationships to the MAC Client. It also shows the terminology used to refer to the frames or packets of data at different layers/sublayers.

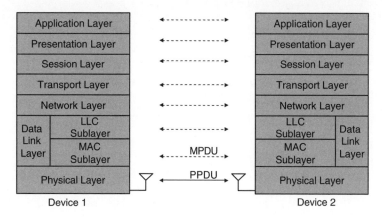

Figure 1.13   ISO/OSI-IEEE 802 reference model

Figure 1.14 shows the following:

- The DME is the layer-independent device manager that controls the device as a whole. It can access all layers as needed. This entity is not part of the OSI model but exists in every implementation. It is considered to be in a different dimension to the OSI model. DME functionality is implementation dependent.
- SAPs are formally defined to indicate the corresponding points of data or control communication between different layers as well as the DME.
- PHY consists of two sublayers, i.e. the PHY Medium-Dependent (PMD) and the Physical Layer Convergence Protocol (PLCP), and a management entity, i.e. the Physical Layer Management Entity (PLME):
  - The PMD sublayer is the entity responsible for the physical transmission and reception of data over the wireless channel to a peer PMD of another device.
  - The PLCP is the sublayer that defines PHY's service interface to the MAC sublayer. It makes the interface to the MAC independent of the PMD.
  - The PLME is responsible for PHY control. It has an interface to the DME called the PLME SAP. Through this SAP, the DME can provide management services to the PLME.
- The MAC sub-Layer Management Entity (MLME) is responsible for MAC control. It has an interface to the DME called the PLME SAP. Through this SAP, the DME can provide management services to the MLME.
- The MAC Service Data Unit (MSDU) is a frame/unit of data that is passed from the MAC Client to the MAC through the MAC SAP.
- The MAC Protocol Data Unit (MPDU) is the frame of data that the MAC protocol prepares for PHY to transmit, or receives from PHY to prepare for the MAC

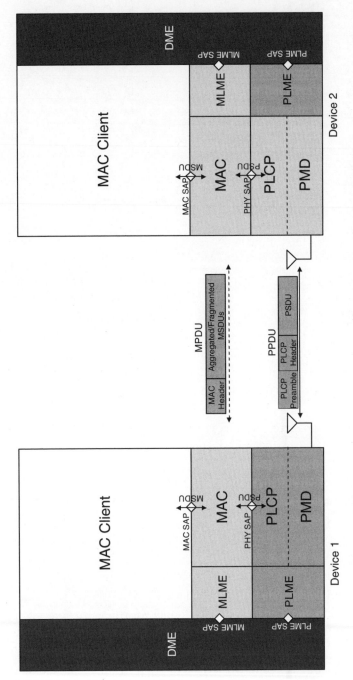

Figure 1.14 MAC and PHY Service Access Points and data frame terminology

Client. As such, it may contain the same data payload as the MSDU plus the MAC header. In many cases, however, the MAC protocol may decide that the payload size received from the MAC SAP is not appropriate for transmission over the medium. In this case, the MAC protocol may aggregate several MSDUs into one MPDU or, vice versa, fragment an MSDU into multiple MPDUs. The receiving MAC sublayer would then de-aggregate or defragment the MPDU to produce the required MSDU for its client.

- The PLCP Service Data Unit (PSDU) is a frame of data that is passed from the MAC sublayer to PHY (specifically to the PLCP sublayer) for transmission over the medium. The PSDU is the same as the MPDU.
- The PLCP Protocol Data Unit (PPDU) is the packet that is transmitted over the Medium. It contains the PSDU, PLCP Header, and PLCP Preamble. In the case of WiMedia PHY, the PLCP Header contains the MAC Header, PHY Header, and FEC to protect the MAC Header and PHY Header. The PLCP Preamble contains packet synchronization and channel estimation sequences.

What is not shown in Figure 1.14 is a MAC Command Data Unit (MCDU), which, unlike the MSDU, does not originate from a MAC Client. Instead, it is generated internally by the MAC protocol as part of its MAC-level command and control with peer MAC sublayers in other devices. Thus, the payloads of Beacon, Command, and Control frames (defined in Section 4.4) of WiMedia MAC are considered MCDUs.

In ECMA-368, Appendix A, there is an additional sublayer called the MUX sublayer. This sublayer has been defined in between the MAC sublayer and its concurrent clients as a way of providing coexistence among multiple protocols running on the WiMedia platform. By adding this sublayer to Figure 1.14 we get Figure 1.15 (shown only for one device).

In this figure, MAC Clients 1–n are simultaneously able to operate with and get service from the MAC sublayer by way of the MUX sublayer, a very simple service that adds a MUX Header to any MUX service data units (data frames from MAC Clients) through the MUX SAP before sending them to the MAC sublayer through the MAC SAP. The MUX Header essentially identifies the protocol (MAC Client) that the data frame belongs to. The peer MUX sublayer in Device 2 recognizes to which protocol an incoming data frame belongs by examining the MUX Header and routes the frame to the appropriate MUX SAP.

The OSI model of network communications is familiar to most communications practitioners from the seven-layer model shown in Figure 1.13. The OSI model specifies layer interactions as well. Figure 1.16 shows the prototypical interactions between OSI layers. A device, Station A, in this case, generates a *request* intended

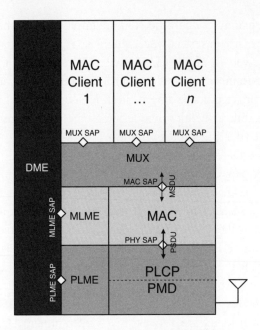

Figure 1.15   MUX sublayer, as in between MAC and its concurrent clients

for Station B. In the figure, the request is generated by Layer $N + 1$ and is intended for Layer $N + 1$ in Station B. The request will progress down through the layers until the PHYs exchange the message between devices. Normally, each layer adds a header to the message as it progresses. The message arrives at Station B, rises up through the layers, each of which strips off any protocol information, until the message arrives at Layer N. Layer N generates an *indication* to Layer $N + 1$,

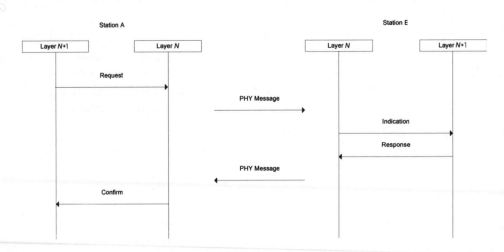

Figure 1.16   OSI protocol interactions

which then processes the request packet, and generates a *response*. The response follows the message path in reverse until it generates a *confirm* at Station A, Layer $N + 1$.

In the WiMedia standard (and many other networking standards) it is possible that a request is simply a message to Layer N and is not intended for Station B. In this case, the request will be acted on by Station A's, Layer N and the confirm is generated by this Layer N without PHY-to-PHY interaction. Similar indication–response pairs may be simply Layer N to Layer $N + 1$ interactions. The WiMedia PHY never responds directly to messages from another WiMedia PHY – it is left to the MAC to handle confirms and responses across the network.

Similar to other standards, ECMA-368 has adopted the use of the words *shall*, *should*, and *may* strictly to mean as follows:

- *shall* refers to mandatory requirements, based on which conformance tests are designed;
- *should* refers to recommendations; implementers are allowed not to follow them;
- *may* gives the permission to act.

Other related words, such as must, will, can, etc., are not used in any way to refer to any of the above meanings. In keeping with the reader-friendly spirit of this book, we will not necessarily adhere to the shall–should–may nomenclature of the standard, however.

The standard also uses the word *octet* to refer to an 8-bit number or sequence. The word byte is avoided, since it has been used in other contexts, albeit infrequently, to refer to numbers or sequences of fewer than 8 bits. Nevertheless, in this book we will interchangeably use the words *byte* and *octet* to mean the same thing, an exactly 8-bit long quantity.

The words *device* and *host* can mean different concepts in different contexts. In this book, we use the lower case *device* to literally mean an appliance, whereas the upper case *Device* is mainly used in the USB/W-USB context to distinguish it from a *Host*. By the same token, the lower case *host* refers to the application that is the ultimate source or destination of data being transmitted. On the other hand, the upper case *Host* is strictly reserved for USB/W-USB master of protocol, controlling the *Devices*.

The words *PHY*, *PHY layer*, and *Physical layer* are all synonymous. So are *MAC*, *MAC sublayer*, and *Medium Access Control*. We may also refer to the MAC and PHY sections of ECMA-368 as WiMedia MAC and WiMedia PHY respectively.

In going from the PHY chapter (Chapter 3) to the MAC chapter (Chapter 4) of this book, the reader will notice that there will be a shift in our terminology. In the

PHY chapter, we will be focusing on *packets*, whereas in the MAC chapter we will be switching to the term *frames*. Although in other standards (e.g. CW-USB) these two terms may be used as if they mean the same, in ECMA-368, *packet* refers to the PPDU, including the PLCP preamble (a non-data-bearing signal). *Frames*, on the other had, are restricted to data units (including any headers) that get passed form layer to layer in the OSI model. Chapter 5, on the other hand, intermixes the terms *packet* and *frame* to be consistent with the terminology of the CW-USB standard.

Throughout the PHY chapter, there are references to the word *symbol* in different contexts. In order to avoid confusion, we adopt the terminology given below in relation to the use of this word throughout the book. The meaning of each of the terms will become clear in the PHY chapter:

- *symbol* – an OFDM symbol plus a zero-postfix;
- *OFDM symbol* – the 128-sample output of the inverse fast Fourier transform (IFFT) in the modulator;
- *data symbol* – a Quadrature Phase Shift Keying (QPSK) or Dual-Carrier Modulation (DCM) symbol used at the input to the IFFT to create an OFDM symbol;
- *modulated symbol* – same as a data symbol.

References

[1] Ecma International Standard ECMA-368, 'High Rate Ultra Wideband PHY and MAC Standard,' 2nd edition. http://www.ecma-international.org/publications/standards/Ecma-368.htm, December 2007.

[2] Barrett T.W., 'History of UltraWideBand (UWB) radar & communications, pioneers and innovators,' Progress in Electromagnetics Symposium 2000 (PIERS2000), Cambridge, MA, July, 2000.

[3] Bluetooth SIG, 'Bluetooth technology to harness the speed of 802.11.' http://www.bluetooth.com/Bluetooth/Press/SIG/BLUETOOTH_TECHNOLOGY_TO_HARNESS_THE_SPEED_OF_80211.htm, February 11, 2008.

[4] Aiello, R. and Shetty, S., Testing raises concerns over 802.11-based high-speed Bluetooth. http://www.wirelessnetdesignline.com/howto/206903929, Wireless Net DesignLine, March 18, 2008.

[5] Ecma International Standard ECMA-369, 'MAC–PHY Interface for ECMA-368,' 2nd edition. http://www.ecma-international.org/publications/standards/Ecma-369.htm, December 2007.

[6] Agere Systems, Inc., et al., 'Wireless Universal Serial Bus Specification,' revision 1.0, May 12, 2005.

[7] ISO/IEC 7498-1:1994, 'Information Technology – Open Systems Interconnection – Basic Reference Model: The Basic Model.'

2

Worldwide Regulations

The regulatory bodies around the world dictate the use of the RF spectrum in their respective jurisdictions. The major regulatory bodies of interest are those that control the major markets for UWB applications. These include:

- Federal Communications Commission (FCC) in the United States;
- Ministry of Internal affairs and Communications (MIC) in Japan;
- European Commission (EC) in the European Community/Union;
- Office of Communications (Ofcom) in the United Kingdom;
- Ministry of Information and Communication (MIC) in Korea;
- Ministry of Information and Industry (MII) in China;

UWB systems are purposefully designed to coexist over a vast frequency spectrum with many other wireless systems, licensed or unlicensed. UWB power emissions are so low that the interference to narrower band systems is expected to be minimal, if any. Nevertheless, the incumbent spectrum holders in the coexistence bands around the world are concerned with having a potential interferer in their licensed band. Worldwide, the incumbent technologies include digital TV, WiMAX, 3rd and 4th generation mobile communication, satellite communication, various governmental communication systems, and airport radar, among others. Thus, it has been a time-consuming and contentious effort for the UWB industry to get the different regulatory bodies around the world to accept the concept of UWB as a coexisting/overlay technology and to allocate proper spectrum and power emission limits for it.

Some regulatory bodies have been pioneering and even aggressive in the adoption of the UWB spectrum, while others have been quite concerned and conservative. As of this writing, the major countries listed above have either allocated or are in the process of finalizing the allocation of the UWB spectrum.

WiMedia UWB: Technology of Choice for Wireless USB and Bluetooth   Ghobad Heidari
© 2008 John Wiley & Sons, Ltd

The UWB regulations will be an evolving one for many years to come. The novelty of overlaying the spectrum for an unlicensed technology over licensed spectra has given many regulators uneasiness. As a result, they will be watching the marketplace and the interaction of the players involved very carefully for the foreseeable future. If UWB is able to peacefully introduce itself into the market without much interference problems to license holders, it is likely that the regulators will be willing to ease up their UWB restrictions.

In the rest of this chapter we will explore the history and status of the regulations in the United States (Section 2.1), Japan (Section 2.2), Korea (Section 2.3), Europe (Section 2.4), and China (Section 2.5). Section 2.6 will then summarize the worldwide picture.

2.1 United States

The FCC was the first regulatory body in the world to allocate unlicensed spectrum for the use of UWB signals. On 14 February 2002, it authorized the use of 7.5 GHz of RF spectrum (3.1–10.6 GHz) for this purpose. Various application areas were contemplated for use in this band, including:

- medical imaging systems;
- communication systems;
- radar and measurement systems (e.g. vehicular radar).

The importance of this allocation is that it allows for the coexistence of the UWB band with other licensed or unlicensed bands in the same spectrum. This is the first time such coexistence has been allowed, especially with such a wide range of spectrum.

The path to FCC approval of the coexistence was not without controversy. In the end, however, the FCC laid down a set of guidelines to avoid undue interference to the incumbent licensed or unlicensed technologies in the band. At the root of these guidelines are the following determinations:

1. *Definition of UWB.* The definition of a UWB signal is that either
   (i) it occupies at least 500 MHz of spectrum,[1] or
   (ii) its 10 dB bandwidth[2] is at least 20% of the size of its center frequency – this is also referred to as 'fractional bandwidth' requirement.

---

[1] There is a proposal to modify this requirement to 450 MHz, since it suits the regulatory requirements around the globe better.

[2] A 10 dB bandwidth is the frequency range between the points on either side of the spectral density curve that are 10 dB below the maximum power density point of the curve.

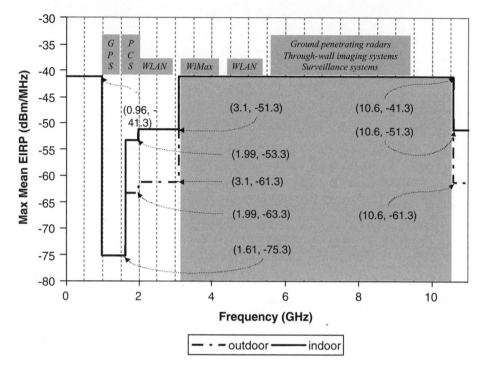

Figure 2.1   UWB power emission limits in the United States

2. *Spectrum allocation for UWB.* The shaded area shown in Figure 2.1 illustrates the allocated frequency band for use of UWB systems in the United States. The range is from 3.1 to 10.6 GHz.

3. *Emission limits.* The power spectral mask depicted in Figure 2.1 is for use in UWB transmitters. The shaded area is the emissions limit for the in-band operation. The rest of the figure indicates the out-of-band emission limits. Note that the average in-band power emission is limited to −41.3 dBm/MHz in terms of equivalent isotropic radiated power (EIRP). This limit was derived from FCC Part 15 rules, where all unintentional radiators are allowed to emit this much unintentional power. This average transmit power limit must be met when measuring with a 1 ms integration time.

   For out-of-band emissions, as seen in Figure 2.1, the emission limits are even lower than those set by Part 15 rules for devices licensed to operate in those bands. This is especially true in the GPS band, where it dips down to −75 dBm/MHz. The out-of-band emission limits are different depending on whether the UWB device is intended to operate indoors or outdoors. Note that any battery-operated/handheld device is considered an outdoors device.

   For peak power, the FCC specifies a limit of 0 dBm in a 50 MHz bandwidth.

This UWB spectrum is already licensed to private and government entities. (Some of the applications in use for this and adjacent bands are named in Figure 2.1.) Despite considerable opposition to such a UWB transmit mask, the FCC granted this ruling, and, by doing so, set in motion the standardization and development of multiple UWB technologies. Although other parts of the world had yet to initiate their UWB regulatory activities, it was a common belief at the time that they would follow the FCC's approach. As further explained in subsequent sections of this chapter, this belief was true, but only to a certain extent.

Originally, the FCC's ruling was based on the assumption that the transmitter would be powered on continuously during the measurement period. As such, a WiMedia device in a hopping mode of operation would have to turn off its frequency hopping (i.e. band sequencing) in order to be tested for FCC certification purposes. This, of course, puts the multiband technologies (including WiMedia UWB) at a substantial disadvantage compared with their single-carrier competitors. In the original hopping mode (TFCs 1–4; see Section 3.3.2), WiMedia's multiband signal hops sequentially over the three bands of a band group in a predefined pattern. As a result, its average power output over the band group is one-third of its output power over any single band.

In response to a petition by certain industry participants (WiMedia), in March 2005, the FCC granted a waiver to allow UWB transmitters that employ frequency hopping, stepping, or sequencing (but not frequency sweeping) to be measured/tested in their 'normal' mode of operation. The waiver is only applicable to indoor and handheld devices that operate in 3.1–5.03 GHz and/or 5.65–10.6 GHz frequency bands. The frequency band for a Microwave Landing System (MLS) and terminal Doppler weather radar fall in the 5.03–5.65 GHz band. Fear of interference with these two very important airport navigation/landing systems caused FCC to block this band out of the waiver ruling.

The waiver allows the WiMedia devices to transmit in their hopping mode at three times the power that they do in their nonhopping mode. This difference in power means that the multiband transmitter can now have a range comparable to that of a single carrier that takes up the same 1.5 GHz of total bandwidth.

A notable omission in the FCC's overall rulings regarding UWB technologies is something called Detection-And-Avoidance[3] (DAA). The FCC never considered such a requirement in any part of the UWB spectrum. However, as we will see in the following sections, some of the other regulatory bodies around the world feel differently.

---

[3] Also referred to as Detect-And-Avoid.

Another difference is the spectrum allocation itself. The FCC is the only regulatory body to allocate as much as 7.5 GHz of spectrum and in a single block. Other regulators fall far short of this, both in size and contiguity.

2.2 Japan

After the United States, Japan was the first to rule on the UWB regulations. Japan's MIC made its first rulings on UWB band allocation back in September 2005, and has revised them somewhat over the years since. The current spectrum allocation and emission limits in Japan are depicted in Figure 2.2, in which the in-band frequency range is shaded (3.4–4.8 GHz and 7.25–10.25 GHz). The FCC's UWB spectrum is also shown in the figure (dotted line) for comparison. The upper limit of the in-band emission is same as in the United States (−41.3 dBm/MHz). However, there are new elements in the MIC's ruling that are not in the FCC's.

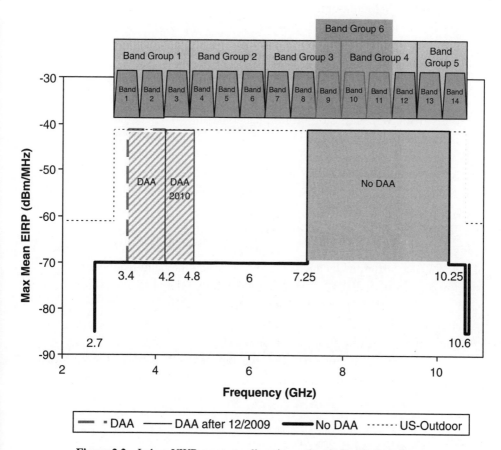

Figure 2.2   Indoor UWB spectrum allocation and emission limits in Japan

First, the regulation allows indoor operation only. The indoor use is regulated by requiring the 'host' device to be mains powered. The 'client' devices are allowed to be battery powered. Of course, in WiMedia's UWB network, there may or may not be a host–client (master–slave) relationship. The WiMedia MAC sublayer, as discussed Chapter 1, is designed to be completely peer-to-peer. No concept of 'host' or 'master' is there to apply to any device. However, when it comes to MAC clients (layers above the MAC sublayer), the network topology may range from peer-to-peer to master-slave. For example, the CW-USB protocol calls for a Host (master) that controls all communication with its Devices (slaves). On the other hand, WLP, WiMedia's adaptation layer to Internet Protocol (IP), keeps the network peer-to-peer. Nevertheless, the main idea from Japan's MIC is clear: keep all UWB communications indoors.

Second, the bands of operation extend from 3.4 to 4.8 GHz and then from 7.25 to 10.25 GHz. Thus, the total spectrum allocated for UWB is much smaller than that in the United States. Also, as seen in Figure 2.2, the allocated bands do not cover all of the WiMedia-defined bands. In fact, Band Groups 1, 2, and 5 are not fully available in Japan.

Third, Japan has added the DAA requirement. As shown in Figure 2.2 in hash-marked shading, there are certain parts of the spectrum that require the use of DAA. Specifically, for the frequency range 3.4–4.2 GHz, Japan's MIC requires DAA. That is, in order for a UWB device to be allowed to operate in this band at the -41.3 dBm/MHz emission limit, it has to be able to detect the presence of any coexisting licensed services (mainly 4G mobile technology) and, if any is present, to be able to avoid its band of operation to prevent any potential interference. Of course, this is easier said than done, and the detailed requirements of DAA (especially detection accuracy and frequency of searches) are not yet well defined. If DAA is not implemented in a UWB device, then it has to adhere to the more stringent emission limit of -70 dBm/MHz in that band, as shown in Figure 2.2.

Figure 2.2 also points to a phased approach to the use of DAA in another part of the UWB spectrum in Japan. The use of DAA in the band from 4.2 to 4.8 GHz is not necessary until 2010.[4] After that date, without the DAA capability, the only band usable for UWB operation may be from 7.25 to 10.25 GHz.

The requirement of DAA in Japan set the stage for other regulatory bodies (other than the United States) to follow suit, as we will see.

Another notable difference between the Japanese UWB regulation and that of the FCC's is in the certification methodology. In particular, in Japan, the conducted

---

[4] In a recent meeting, International Telecommunication Union (ITU) decided to limit band allocation for 4G/3G/WiMax to below 4.2 GHz. Hence, there is a reasonable chance that regulatory bodies will relax 4.2–4.8 GHz requirements in the near future.

(not radiated) emission levels are checked against the spectrum mask levels of Figure 2.2. That is, unlike in the United States, where the antenna is considered an integral part of the transmitter, in Japan, a UWB transmitter under test is expected to have a wired connection between its RF output and the test equipment. This results in major differences, for example, on the levels of spurious emissions measured for US certification versus for Japan.

The UWB industry is still trying to lobby Japan's MIC to reduce some of its severe restrictions to better fit the UWB market requirements. In fact, the hope is to influence Japan's MIC to add sufficient allocation to the UWB spectrum to allow Band Group 3 to be operable in Japan with DAA. However, predicting that this may not occur, WiMedia has already developed the new Band Group 6 that would be better suited for use in Japan, as well as other regions of the world.

The overlay of the WiMedia band groups over the UWB spectrum in Figure 2.2 clearly shows that the only usable band groups that are not hampered by DAA are Band Groups 4 and 6. As we will see later, Band Group 6 turns out to be the most useful DAA-free band group for worldwide operation. As is expected, compared with the lower frequency band groups, the propagation range in this band group is not as desirable.[5] Therefore, there is an incentive for manufacturers to develop devices with DAA. However, the extra complexity of DAA is bound to drive the price and power consumption of such devices higher.

Note how the mitigation-required portion of the allocated UWB spectrum (the hatch-pattern shaded area) does not even cover Band Group 1. This is a major problem for WiMedia devices as it handicaps them even if they have a perfectly compliant interference mitigation technique implemented. They cannot use Band Group 1 to its fullest extent. Instead, they can only use two of the three bands in that band group.

The Japanese UWB regulation was the first one that put the WiMedia Alliance in a predicament regarding the use of Band Group 1. As indicated in Figure 2.2, Band Group 1 is not fully usable for UWB under Japanese regulations. Even with DAA, part of Band 1 is not within the allocated spectrum. Thus, while the first version of the WiMedia PHY specification made Band Group 1 a mandatory feature of every manufactured WiMedia device, the use of that very band group is disallowed in Japan. The unavailability of this band group as a universal band group forced the Bluetooth SIG (which was, at the time, contemplating the use of WiMedia UWB for its Alternate MAC-PHY technology) to require specifically that Band Group 1 not be mandatory. Eventually, the WiMedia Alliance decided to remove the mandatory requirement of Band Group 1, as is now the case in the latest revision of the PHY specification [1].

---

[5] The higher the carrier frequency is, the higher the propagation loss is, and thus the lower the signal range is.

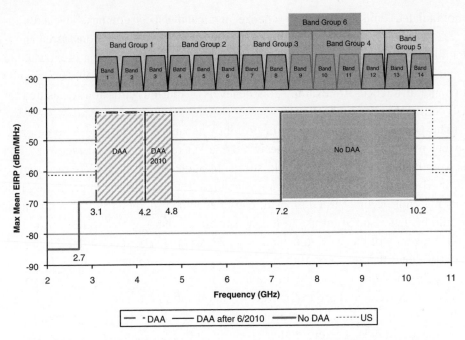

Figure 2.3 Outdoor (portable) UWB spectrum allocation and emission limits in Korea

2.3 Korea

In July 2006, Korea's MIC allocated UWB spectrum similar to that in Japan. The current status of this allocation is as shown in Figure 2.3. One major difference is that, unlike its Japanese counterpart, the Korean UWB spectrum starts at 3.1 GHz and ends at 10.2 GHz. This allows Band Groups 1 and 4 to be fully usable in Korea. However, similar to Japan, Korea has put interference mitigation (DAA) requirements on the low end of the spectrum. From 3.1 to 4.2 GHz, DAA is required. From 4.2 to 4.8 GHz, DAA is waived until June 2010.[6] From 7.2 to 10.2 GHz, no interference mitigation is ever required.

Although DAA is mentioned as the method of interference mitigation, the actual regulation calls for a UWB device operating in the 3.1–4.8 GHz band to meet one of the following options:

1. transmit less than −70 dBm/MHz;
2. transmit less than −41.3 dBm/MHz with transmission period of no more than 5 ms after each nontransmission period of at least 1 s;

---

[6] In a recent meeting, the ITU decided to limit band allocation for 4G/3G/WiMax to below 4.2 GHz. Hence, there is a good chance that regulatory bodies will relax 4.2–4.8 GHz requirements in the near future.

3. deploy a DAA algorithm such that the emission power is reduced to the −70 dBm/MHz level (or completely stopped) within 2 s of detecting a signal stronger than −80 dBm/Hz.

Figure 2.3 also superimposes the WiMedia band groups over the Korean UWB spectrum. Clearly, the only band groups that are not encumbered by interference mitigation are Band Groups 4 and 6.

2.4 Europe

Under pressure to act on adopting a standard for UWB to enable the use of UWB products in the European Union (EU), the European Commission (the Commission) released a policy document [2] on 21 February 2007 in favor of allocating spectrum for UWB. Although it did not set all the details of the regulations, it did acknowledge the need for allowing UWB to provide the foundation for a large set of applications (communication, location, imaging, medical, etc.). By allowing a harmonization of the spectrum across the European Communities (EC), the Commission hoped to create a single market with a large economy of scale. Of course, concern about the UWB grey market (devices with regulatory specifications inappropriate for the EU from other parts of the world finding their way into the EU market) was another impetus for the EC to take action. The Commission's decisions are important, since they can considerably influence the RF spectrum regulations not only within the EC, but also outside it. Countries like China, Korea, and even Japan have been looking to the EC to determine how to finalize their own regulations. It was not desirable for the EC to be the last major market to decide on UWB.

The Commission made it very clear in this policy document that the incumbent, licensed services[7] must be protected against undue interference from UWB devices. It clearly supported the notion of adding some sort of a DAA or Low-Duty-Cycle mechanism for UWB devices to mitigate any potential interference to the victim services. A Low-Duty-Cycle mechanism was defined by the Commission to be one that restricts the transmission of all UWB signals to

1. be less than 5 % of the time in each second of time;
2. be less than 0.5 % of the time in each hour;
3. have each transmission being no longer than 5 ms in duration.

---

[7] The incumbent services of interest in the EC consist of WiMax and 4G mobile services for indoor operation, and military radar for outdoor operation.

However, the Commission left it to the European Conference of Postal and Telecommunications Administrations (CEPT) and ETSI to determine the details of the mitigation requirements.

It is worth mentioning that, at the time the Commission issued this decision, intensive technical studies mandated by the Commission were being conducted by the Electronic Communications Committee (ECC) of CEPT to review the relevance of the conservative protection limits and for defining appropriate mitigation techniques for the protection of the three incumbent services [3]:

- Civil Aeronautical radars, in 2.7–3.1 GHz
- Military radars, in 3.1–3.4 GHz, and in 8.5–9 GHz.
- Broadband Wireless Service terminals in 3.4–4.2 GHz (including WiMAX in 3.4–3.8 GHz)

As of this writing, the results of these studies have just been submitted for consideration by the ECC, with the recommendation to review the protection limits in these bands, and to adopt mitigation techniques (DAA and LDC) between 3.1–4.2 GHz and 8.5–9 GHz. The results show that, in the frequency band 4.2–4.8 GHz, no mitigation technique may be necessary as the operation of UWB devices at −41.3 dBm/MHz was considered of low risk of interference with existing outdoor services deployed in this band. A revised ECC decision has been drafted for public consultation, including these amendments; however, the final approval has not yet been issued.

Based on the current European Commission decision [2], the maximum mean EIRP for UWB transmissions are set as shown in Figure 2.4. The DAA-mandatory portions of the allocated frequency bands are also identified in Figure 2.4, excluding the potential additions to the DAA bands upon the completion of the aforementioned ECC studies. As it currently stands, the only unencumbered part of the spectrum is from 6–8.5 GHz. From 3.4–4.2 GHz, there has to be a mitigation technique acceptable to CEPT/ETSI before high enough emissions (−41.3 dBm/MHz) are allowed for viable WiMedia UWB operation. Note that the victim services in this band are mainly Aeronautical, some Military Radar, and indoor WiMAX terminals.

Provided the aforementioned DAA technical studies successfully prove the viability of allowing DAA-enabled UWB devices 3.1–3.4 GHz and 8.5–9 GHz bands, Figure 2.4 will transform to Figure 2.5.

The out-of-band emission limits of Figure 2.4 are extremely restrictive, due to the very conservative protection limits considered in this initial EU decision. As shown in Figure 2.5, these limits may be revised based on the results of the technical studies. Nevertheless, these out of band emission levels will still remain lower than in the US. Therefore, dedicated filters in the RF frontend of the UWB devices may be required to comply with such levels.

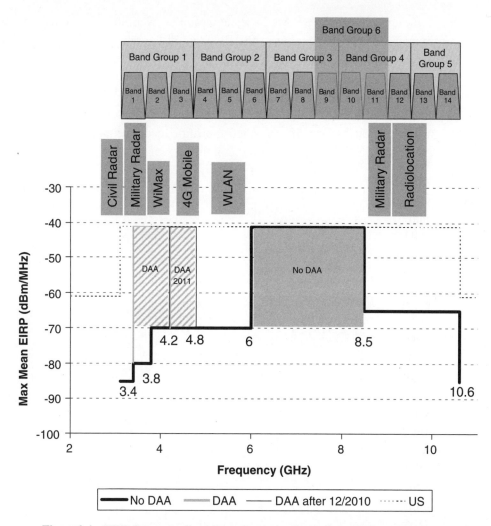

Figure 2.4 UWB Spectrum allocation and emission limits set by ECC on Feb 21, 2007

Based on the revised technical studies, from 4.2–4.8 GHz, no mitigation technique is required. However, the deployment of future Radiocommunication systems between 4.2 and 4.8 GHz is still a possibility, and a further revision of UWB regulatory framework will certainly be require.[8]

From 6 GHz–8.5 GHz, no interference mitigation is ever required.

Currenlty no operation above 8.5 GHz is allowed. However, from 8.5–9 GHz, DAA-enabled UWB operation may be allowed if the draft ECC report [3] is approved.

---

[8] In a recent meeting, ITU decided to limit band allocation for 4G/3G/WiMAX to below 4.2 GHz. Hence, there is a good chance that regulatory bodies will relax 4.2-4.8 GHz requirements in near future.

Figure 2.4 superimposes the WiMedia Band Groups as well. Recall that this figure is based on the current standing of the EU Commission on UWB emissions. Note how the allocated UWB spectrum does not cover Band Groups 1 or 6. This is a major problem for WiMedia devices as it handicaps them even if they have a perfectly compliant interference mitigation technique implemented. They cannot use Band Group 1 to its fullest extent. Instead, they can only use two of the three bands in that Band Group. However, if and when the draft recommendations of ECC are approved within ECC and later by the Commission, as shown in Figure 2.5, Band Groups 1 and 6 will be usable in the EU.

ETSI regulation on UWB is now formally published and hence certification of UWB products can be formally started for European market for 4.2–4.8 GHz (till sunset date) and 6–8.5 GHz now. Also the DAA regulatory requirements were

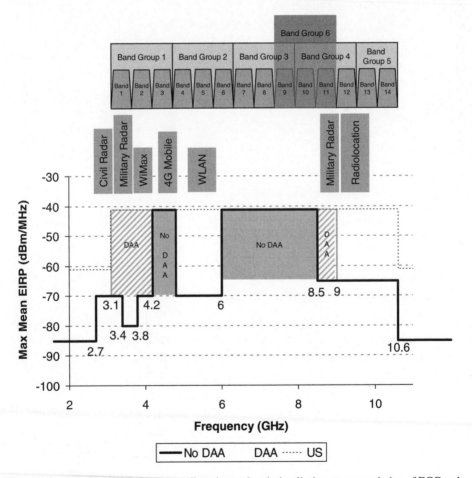

Figure 2.5 Potential UWB Spectrum allocation and emission limits upon completion of ECC technical studies on DAA

concluded at the ECC TG3 meeting of 19 February 2008. As a result a public consultation on DAA formally started in early March 2008 in Europe. For now, based on current progress, ETSI plans to complete the revision of the European UWB regulation for inclusion of DAA by the end of 2008 with a target publishing date of the final regulation in the second quarter of 2009.

Compared to the Japanese and US allocation of spectrum (Figure 2.2), the Commission has allocated far less spectrum for UWB. In fact, it does not extend far enough to cover WiMedia's Band Group 6. This is also a problem for WiMedia since Band Group 6 is its only unencumbered Band Group that is hoped to become universally acceptable around the world.[9] The reason for this shortcoming in the Commission's allocated spectrum is the concern over the interference with NATO's radars operating in those frequencies.[10]

The report [2] also mandates each member state of the EC to adopt UWB within six months of the date of its publication. The use of UWB devices in in-vehicle applications is not recommended in this initial decision. The potential interference from such operations have also been studied by ECC since the publication of this decision. At the time of this writing, recommendations for adopting transmit power control as a mitigation technique for in-vehicle applications have been submitted for consideration to EC and a revised decision has been issued [4].

In summary, there are many areas of potential improvement to the allocated spectrum and the WiMedia member companies are lobbying the Commission/CEPT/ETSI to make changes such as:

(a) provide for DAA operation on the low end of the band to allow Band Group 1 to be fully usable;

(b) increase the out-of-band emission limits on the low end of the spectrum;

(c) increase the higher edge of the allocated spectrum to 9 GHz (to enable Band Group 6);[11]

(d) clarify and simplify the DAA requirements.

As such, the process of spectrum allocation and emission requirements in Europe is not finalized yet. The UWB Industry is contributing intensively to ECC towards defining the regulatory requirements with adequate protection limits and optimum mitigation techniques, hoping to open up more of the spectrum to allow Band Group 1 and Band Group 6 to be fully utilized.

---

[9] Band Group 6 was specifically added to the WiMedia specifications to provide such a universally-unencumbered Band Group.

[10] Mainly in Germany.

[11] In 2006, WiMedia ran some real-time test measurements to convince the NATO authorities that there should not be any harmful interference from UWB operation in this region of the spectrum near NATO facilities.

2.5 China

Figure 2.6 depicts the UWB spectrum allocation and emission limits in China, as of this writing. The regulations are not finalized yet, but the familiar theme of interference mitigation requirement in the lower end of the band persists here as well. For now, from 4.2 to 4.8 GHz, DAA is required. Devices using this band are restricted to indoor operation only. From 6 to 9 GHz, no interference mitigation is expected and no restriction is put on the indoor or outdoor operation.

2.6 Summary

Although there is still some room for improvement in the global spectrum allocation, UWB device manufacturers are now able to deploy their products in major markets around the world. There has been a slow but steady overall progress in worldwide UWB regulations so far. After approval of the UWB spectrum in the United States by the FCC, regulatory bodies in Japan, Korea, Europe, and China followed suit in allocating UWB spectra in their respective regions. Figure 2.7 illustrates the relative spectrum allocation in these markets as of this writing (excluding the potential relaxation of EU regulations based on the ECC technical

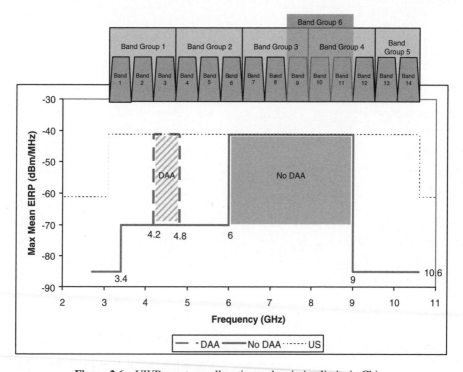

Figure 2.6   UWB spectrum allocation and emission limits in China

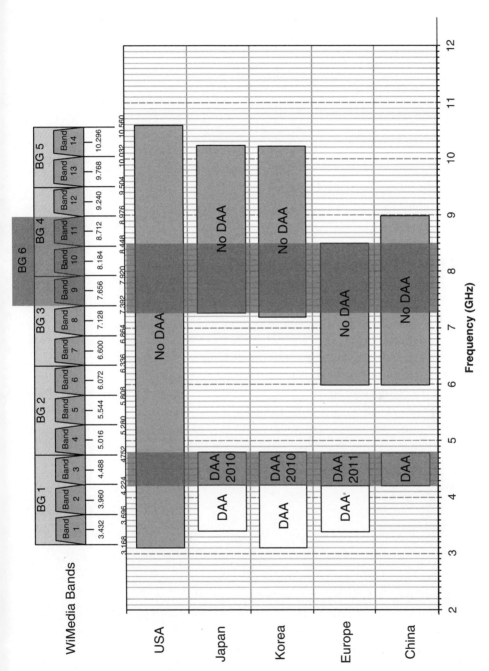

Figure 2.7 Summary of current worldwide UWB spectrum allocations

studies on DDA that are currently underway, as discussed in Section 2.4). Obviously, the US market is the most open to UWB, with no interference mitigation requirements and with the widest contiguous frequency band allocated. The out-of-band emission limits in the United States are also the least stringent. On the other hand, the rest of the world seems to follow a pattern of:

1. allocating two disjoint frequency bands;
2. requiring interference mitigation (DAA) in the lower band.

In particular, the two bands of 4.2–4.8 GHz (with a sunset date) and 7.25–8.5 GHz are the only portions of the RF spectrum that are currently allocated for UWB across all the major regions. These are shown as two shaded strips crossing all regions in Figure 2.7. The left strip requires DAA, whereas the right one does not. Note that neither of the two frequency ranges covers an entire WiMedia band group. In the lower frequency range, only WiMedia Band 3 of Band Group 1 fits, while in the upper range, Bands 9 and 10 of Band Group 6 fit. Of course, WiMedia specifications allow devices to operate in a single band (Fixed Frequency Interleave (FFI)) and even in two bands (Time–Frequency Interleave with two bands (TFI2)). However, the optimal coverage range and data rate are achieved when a whole band group is utilized.

On the bright side, there are two hopeful developments for WiMedia and its members to look forward to. First, the regulatory bodies are far from finalizing their UWB regulations. Currently, ECC is considering recommending creating additional DDA-enabled UWB bands in 3.1–3.4 GHz and 8.5–9 GHz. It is reasonable to expect that Japan and China will follow the EU's lead in opening the entire 3.1–4.2 GHz band, subject to DAA.

These would open up Band Groups 1 and 6 to be universally usable, subject to DDA. Of course the more desirable outcome would be to have EU allow non-DDA operation in the 8.5–9 GHz range. If this happens, Band Group 6 would be universally available without DDA, making it quite popular among manufactures of inexpensive devices.

Second, even without any regulatory harmonization, one can rely on the latest advancements in CMOS RF technology to enable UWB devices to operate universally without the need for multiple/multimode RF frontends. Some companies are building multiband RFICs that can handle at least two WiMedia band groups. As of now, there is at least one company[12] which has developed a CMOS RFIC with a bandwidth wide enough to cover the entire UWB spectrum (3.1–10.6 GHz). In the

---

[12] Wilinx Corporation.

future, there will likely be more such RFIC developments, making the universality of the UWB devices a reality even without a harmonized UWB spectrum.

A device built with a wideband or multiband RF frontend can easily switch its frequency range of operation depending on the region of the world it is in. All that is needed is for the device to be able to recognize where in the world it is at any given time. Location awareness does not come easily, especially if one thinks of portable devices that travelers can easily transport across international boundaries. However, the WiMedia specifications already include provisions for devices to announce/distribute their location (regulatory domain) information amongst themselves, making the location determination task simpler. GPS and other location-identifying subsystems in the portable devices can also be helpful.

In the coming years, and with the proliferation of the UWB devices around the world, the question of whether the interference concerns are real or imaginary will be proven in the field. The regulatory bodies will start using real field data (as opposed to conservative theoretical analyses) to decide how much protection that licensed services practically require from UWB devices. Once the dust settles, expect to see

- some of the excessive restrictions to be lifted;
- the DAA regulations to be detailed and refined as necessary;
- additional spectrum-harmonizing adjustments take place in different countries;
- UWB technology advancements to meet the regulatory requirements better.

References

[1] Ecma International Standard ECMA-368, 'High Rate Ultra Wideband PHY and MAC Standard,' 2nd edition. http://www.ecma-international.org/publications/standards/Ecma-368.htm, December 2007.
[2] Commission of the European Communities, 'Commission Decision of 21/II/2007 on allowing the use of the radio spectrum for equipment using ultra-wideband technology in a harmonised manner in the Community,' Brussels, February 21, 2007.
[3] Draft ECC Report 120, "ECC Report on Technical Requirements for UWB DAA (Detect and Avoid) Devices to Ensure the Protection of Radiolocation in the Bands 3.1–3.4 GHz and 8.5–9 GHz and BWA Terminals in the Band 3.4–4.2 GHz," Electronic Communications Committee (ECC) within the European Conference of Postal and Telecommunications Administration (CEPT), March 2008.
[4] Amended ECC/DEC/(06)04, "ECC Decision of 24 March 2006 amended 6 July 2007 at Constanta on the harmonized conditions for devices using Ultra-Wideband (UWB) technology in bands below 10.6 GHz," Electronic Communications Committee (ECC), amended 6 July 2007.

3

Physical Layer

Robert T. Short, PhD

The Physical Layer (PHY) is the lowest layer in the OSI reference model (see Figure 1.13). The PHY is the portion of the system that converts data into waveforms suitable for transmission across the physical medium. In the case of UWB, the medium is simply the air between the devices along with the frequency spectrum allocated by the regulatory bodies. The waveform selected by the WiMedia committee is a frequency-hopped OFDM radio signal. In this chapter, we will describe both the waveform and related issues in detail.

We will present a very high-level overview of the PHY in this chapter, introducing some of the basic concepts and terminologies. We then explore each of the major sections in turn. We use the terminology of the WiMedia standard, to simplify the reader's understanding of the standard. We will introduce the terminology here, but discuss and redefine as we proceed.

Throughout this chapter, we will adhere to the terminology of Section 1.7.

Figure 3.1 shows the structure of the PHY at the most basic level. The PHY simply translates data that is received from the MAC to waveforms that may be transmitted over the UWB radio channel. The purpose of this chapter is to describe that translation in simple but complete terms.

We will describe the entire PHY architecture in some detail, but it is important to note that the WiMedia standard does not require any structure or architecture; it only specifies the air interface. All of the architectural detail found in the standard or this chapter is simply to provide a context in which to describe the required PHY functionality. Most implementers, however, seem to find it convenient to nominally follow the structure described in the standard. Similarly, the standard rarely

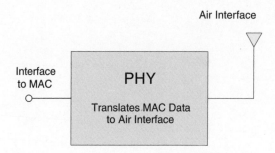

Figure 3.1   Basic block diagram of the PHY layer

requires any particular implementation, either for the receiver or the transmitter. The emphasis, both in this chapter and in the standard, is on the transmitter.

The PHY is partitioned into several layers, as shown in Figure 3.2. In this section we will briefly introduce the layers and interfaces, and we will progress to very detailed descriptions as we proceed through the chapter. There are two 'upper' interfaces: the management interface and the data interface. The management interface, i.e. the PLME SAP, as its definition suggests, is a side channel into the PHY for the purpose of controlling that layer's operation. The data interface, i.e. the PHY SAP, is the path through which data is passed between the MAC and the PHY.

We will defer any discussion of the WiMedia PHY management sublayer to a later section. At this time we are concerned with the data path. Data to be transmitted is passed to the PHY via the PHY SAP and consists of a data segment (the PSDU), a MAC header, and some control information (data length, etc.). The PLCP sublayer presents the PHY SAP interface to the MAC in a form that is independent of the air interface and formats the data into a digital waveform, i.e. into a form that is dependent on the air interface. The result of the PLCP operation

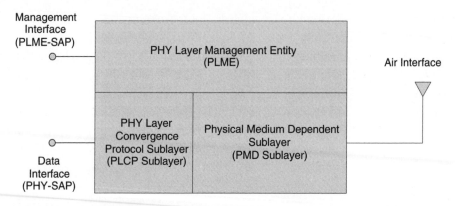

Figure 3.2   PHY sublayers and interfaces

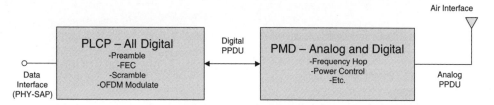

Figure 3.3   PHY data path

is passed to the PMD, where it is converted to an analog waveform. The 'lower' interface, the air interface, is the antenna. Of course, received data follows the reverse path.

Note that the WiMedia standard does not specifically define the functions assigned to the PMD sublayer. It does, however, specify that certain functions are in the PLCP; so, for the purposes of this document, the PMD is assumed to be everything not specifically included in the PLCP by the standard. In fact, the standard is not entirely clear or consistent within itself, nor is it entirely consistent with other published literature as to the form and function of the various sublayers. As noted before, the architectural boundaries are for reference, so it is only important to define the boundaries for exposition.

The PLCP sublayer transforms the MAC header and data (PSDU) into a PPDU. The PPDU includes the header and data, but after error correction, scrambling, and other processing has been applied (or removed during the receive process; see Figure 3.3). The PSDU is a sequence of bits, i.e. a sequence of 1s and 0s. The PPDU at the output of the PLCP is a digital waveform, or a sequence of finite-precision numbers. The digital PPDU is converted to an analog PPDU by the PMD and transmitted over the air. Note that the OSI model defines the PPDU as the actual bits carried by the waveform, rather than the waveform itself, but the WiMedia standard clearly refers to the PPDU as a waveform.

It will be useful to further partition the PLCP into an outer sublayer and an inner sublayer. The outer sublayer includes scrambling, error correction, and other functions that operate on bits, while the inner sublayer converts bits to finite-precision waveforms using OFDM. OFDM waveforms are composed of a sequence of distinct entities known as *symbols*[1] and each symbol is 312.5 ns long in the WiMedia waveform. At the end of each symbol, the PHY layer may hop to a new center frequency. Again, we will describe all of this in considerable detail in the following sections, but it is important to understand that the frequency hopping occurs each symbol or every 312.5 ns in the WiMedia PHY.

---

[1] As explained in Section 1.7, throughout this book, we will be consistently using the term *symbol* in different contexts.

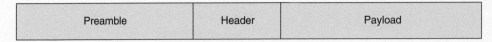

Figure 3.4 PHY packet structure

The PMD translates the digital PPDU into analog waveforms that are transmitted (and, of course, received) by the antenna. The PMD includes the data converters, frequency hopping, RF up-conversion, gain control, etc. Put another way, the PLCP sublayer is related to the digital baseband in the PHY and the PMD sublayer refers to the analog-to-digital conversion (ADC)/digital-to-analog conversion (DAC) and the RF portion of the PHY.

The WiMedia system is a packet radio system, i.e. all information is transmitted in packets. A packet generated by the PHY is a waveform that consists of three major parts, as shown in Figure 3.4. The preamble conveys no information, but is a waveform of predetermined length and structure that is used to allow the receiver to detect the existence of the packet and to estimate the parameters needed for accurate demodulation. The header is a short burst of data; part of it is provided by/for the MAC sublayer and part of it is constructed by the PHY with information such as data rate, scrambling seed, etc. Of course, the payload contains the information that one device is conveying to another.

The details of the PLCP and the PMD themselves are strongly dependent on the final waveforms presented at the antenna.

Section 3.1 gives a quick summary of the PHY features and capabilities. Section 3.2 is a combination of a tutorial on OFDM and a detailed description of the WiMedia OFDM waveform structure. We then proceed to build the WiMedia PHY standard from the bottom up, which is precisely the opposite of the approach taken in the standard. We discuss the RF in Section 3.3, then FEC and related functions in Section 3.4, and finally put together an entire packet in Section 3.5. PHY performance requirements are discussed in Section 3.6, followed by a look at the PHY responsibilities for range measurement in Section 3.7. We will then return to Figures 3.2 and 3.3 in much more detail, expanding on each of the interfaces, the data paths, and the sublayers in Section 3.8.

3.1 Feature Summary

3.1.1 Packet Radio

The WiMedia PHY is a packet radio system. Information is transmitted in chunks. The standard includes a standard mode for transmitting individual packets and a mode for transmitting bursts of packets for greater throughput. Each packet includes a preamble for acquisition and parameter estimation, a header that conveys basic modulation and packet information, and a payload section that carries data.

3.1.2 MB-OFDM

The PHY uses OFDM to transmit information: 128 subcarriers are used, 100 of which carry data, 12 of which are pilots, and 10 of which are reserved as guards. The remaining subcarriers are zeros. The WiMedia OFDM uses a zero-postfix or zero-padded suffix instead of a cyclic prefix to mitigate the effects of multipath.

The subcarriers are modulated using QPSK or a new technique called DCM, a variation on 16 quadrature amplitude modulation (QAM). Higher order modulation types, such as 64 QAM, are not used because of the low power required by the regulatory agencies.

The PHY requires the ability to frequency hop over multiple bands of frequency. The center frequency is selected from a set of three available center frequencies or bands, which are in turn selected from a set of band groups. the band groups span the entire 7.5 GHz of the UWB spectrum.

The WiMedia PHY uses a technique called spreading in which data is placed on multiple OFDM subcarriers to gain diversity and thereby improve performance in fading channels.

3.1.3 Error Correction and Variable Data Rate

The WiMedia PHY uses a convolutional code with puncturing. The combination of spreading and variable code rates gives the PHY a selection of data rates allowing for very high data rates in good channels over short distances and good performance in poor channels and over longer distances. The available payload data rates are 53.3, 80, 106.7, 160, 200, 320, 400 and 480 Mbits/s. In addition to error correction, interleaving and scrambling are an integral part of the packet structure.

3.2 WiMedia OFDM

As discussed before, except for the interface presented to the MAC, the processing performed by the PHY is strongly dependent on the waveforms presented at the air interface. This section is a tutorial on the OFDM waveform structure. We will motivate the use and structure of the OFDM modulation, describe the basics of OFDM, and discuss some of the interesting features that are unique to the WiMedia system.

3.2.1 Frequency-selective Fading

Recall that the world's regulatory agencies are defining a UWB waveform as one that has a minimum of 500 MHz of bandwidth. If we attempt to transmit a traditional single-carrier waveform through an indoor channel, then in many situations, perhaps even most, severe distortion of the signal will occur. In Figure 3.5, a binary phase-shift keying (BPSK) signal is modulated to 4 GHz, transmitted through a typical UWB channel, and then down-converted back to baseband. The pulse distortion is severe, and since the resulting pulse is much longer than the transmitted pulse, significant intersymbol interference (ISI) will occur. Figure 3.6 shows the same signal in the frequency domain. The deep notches in the spectrum, along with the associated rapid phase changes, create the pulse distortion and is a phenomenon known as frequency-selective fading. This type of fading is a significant problem even in narrowband waveforms, but the wide bandwidths associated with UWB make the problem particularly severe.

The traditional method for dealing with this type of ISI is to use a channel equalizer. Such equalizers are well known and understood, but require adaptive signal processing to be effective. Adaptive signal processing requires time to converge and special preamble structures are often needed to allow for convergence in a packet radio environment. The equalizers are often very large, and if the equalization process fails on a packet, then the entire packet is lost. As we will see, OFDM provides an alternative to the equalizer structure. In particular, it will replace the

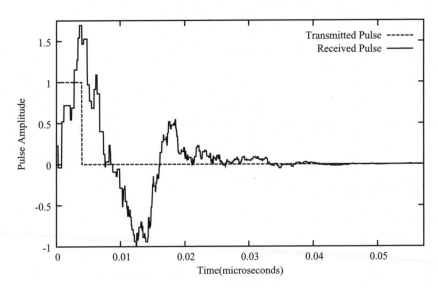

Figure 3.5 Transmit and receive pulses for a 500 MHz single-carrier waveform over a UWB channel

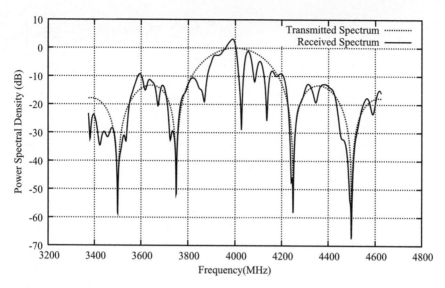

Figure 3.6 Transmit and receive spectrum for a 500 MHz single-carrier waveform over a UWB channel

equalizer with a different set of complex structures. OFDM is not necessarily superior to single-carrier modems, but does offer a different set of tradeoffs.

Observing the effect of lengthening the pulse motivates the OFDM approach. Figures 3.7–3.10 show the impact of the channel on a BPSK signal transmitted as before. Figures 3.7 and 3.8 show a pulse that has been increased in length by a factor of 10 over the 500 MHz waveform (reducing the bandwidth to 50 MHz).

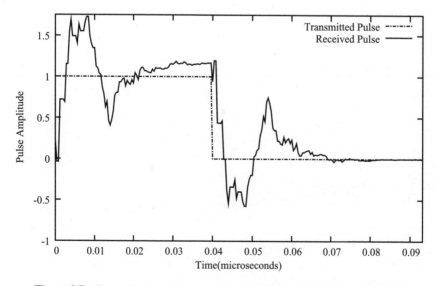

Figure 3.7 Transmit and receive pulses for a 50 MHz single-carrier waveform

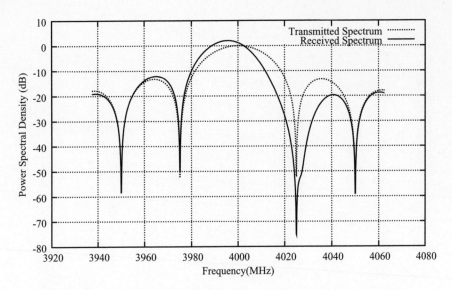

Figure 3.8   Transmit and receive spectra for a 50 MHz single-carrier waveform

Figures 3.9 and 3.10 show the results of the pulse stretched by a factor of 100; that is, a pulse with a bandwidth of 5 MHz. Clearly, as the length of the pulse increases, the distortion reduces and the relative amount of ISI decreases. Of course, in the process, the amount of information that may be carried is also reduced; and since the bandwidth is less than 500 MHz, such systems are no longer UWB waveforms.

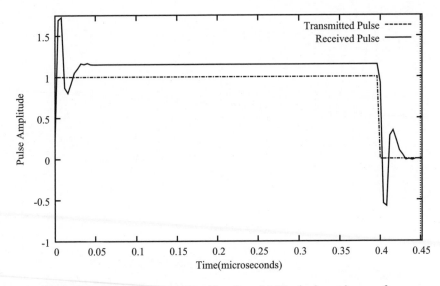

Figure 3.9   Transmit and receive pulses for a 5 MHz single-carrier waveform

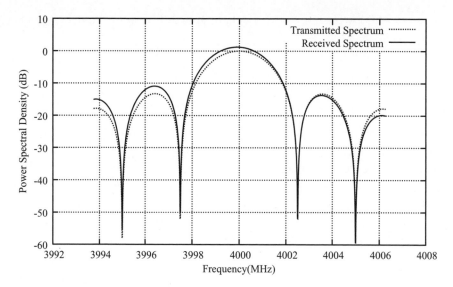

Figure 3.10 Transmit and receive spectra for a 5 MHz single-carrier waveform

If we add more carriers, however, we are able to increase both the bandwidth and the information-carrying capacity of the system. Figure 3.11 shows a system in which five 5 MHz pulses are modulated onto *subcarriers* spaced 100 MHz apart. Clearly, the bandwidth is nominally 500 MHz and the information capacity is 11 times the single-carrier version. Since the pulses are all 100 times longer than a

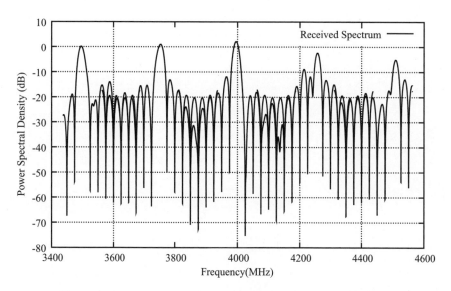

Figure 3.11 Transmit and receive spectra for a multicarrier waveform

single-carrier 500 MHz pulse, the effects of signal distortion and ISI are reduced on each subcarrier.

Just as clearly, it is important to pack as many subcarriers as possible into the system in order to maximize the information-carrying capacity of the multicarrier system. We will decrease the spacing between the subcarriers to the smallest possible value, but will require that the subcarriers do not interfere with each other in the demodulation process. Subcarriers that do not interfere with each other are *orthogonal* subcarriers, and the minimum spacing to ensure orthogonality is the bandwidth of the pulse; that is, for the 5 MHz pulses used in this example, the minimum spacing is 5 MHz. With the spacing reduced to the minimum, we have an OFDM waveform, with the transmit spectrum shown in Figure 3.12. In our example, we have stretched the pulse out by a factor of 100 when compared with a single-carrier system, but packed 100 subcarriers into the same spectrum. Each of the subcarriers is subject to amplitude distortion, but the phase distortion per subcarrier is relatively small; in other words, we have reduced the frequency-selective fading problem to a collection of flat fading ones. The collection of modulated pulses for a single pulse time is an *OFDM symbol*.

It might seem that further lengthening the pulse and adding more carriers would provide even greater benefit. However, there are a number of reasons to limit the process. First, the channels do vary with time. Any motion in the room will cause a change, and it is not generally desirable to adapt the system during the reception of an OFDM symbol, so the pulses need to be significantly shorter than the channel variability. Second, adding carriers also increases the system complexity, so we

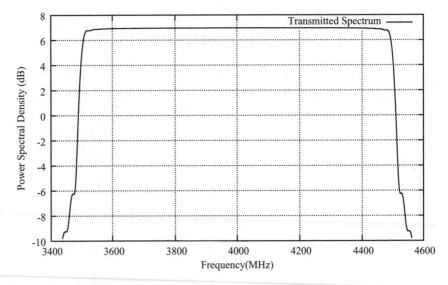

Figure 3.12   Transmit spectrum for an OFDM waveform

are further driven to add only as many as needed. There are other reasons that are beyond the scope of this text.

3.2.2 Mathematical Framework

This text is not intended to be a detailed mathematical treatise on OFDM, but it will be necessary to have a basic framework in which to describe our system. This section provides a simplified mathematical view of OFDM.

The complexity of OFDM waveforms generally precludes an analog implementation, so we assume a digital system for the generation and reception of our waveforms. To that end, our mathematics will assume sampled waveforms. In general, sampled waveforms require a sample rate, which we will denote f_s and an associated sample interval $\Delta t = 1/f_s$. The WiMedia PHY uses $f_s = 528$ Msamples/s. We will describe a single unit-energy rectangular pulse that is N samples long as

$$p_N[n] = \begin{cases} 1/\sqrt{N} & 0 \le n < N \\ 0 & \text{otherwise} \end{cases}$$

Each pulse will eventually be modulated up to a final subcarrier frequency, but we will pre-modulate each one to an intermediate frequency

$$\psi_k[n] = e^{i(2\pi/N)kn} p_N[n]$$

where k is the subcarrier number (any integer, positive, negative, or zero). Most OFDM systems use linear modulation, usually QAM, and $X[k]$ is a complex number representing the modulated *data symbol* (or simply modulated symbol) for the subcarrier intermediate frequency $(2\pi/N)k$. We combine $2K + 1$ modulated pulses additively into the mth *OFDM symbol*

$$s_m[n] = \sum_{k=-K}^{K} X_m[k]\psi_k[n] = \sum_{k=-K}^{K} X_m[k]e^{i(2\pi/N)kn} p_N[n]$$

where an OFDM symbol is exactly N samples long. We then transmit M OFDM symbols sequentially, so that the equation describing the basic OFDM waveform discussed so far is

$$s[n] = \sum_{m=0}^{M-1} s_m[n - mN_s] = \sum_{m=0}^{M-1} \sum_{k=-K}^{K} X_m[k]e^{i(2\pi/N)k(n-mN)} p_N[n - mN]$$

The key to practical implementation of OFDM is to synthesize and demodulate the waveforms with a fast Fourier transform (FFT) algorithm. The FFT is an efficient implementation of a discrete Fourier transform (DFT). One definition of the DFT pair consists of the synthesis equation or inverse DFT (IDFT)

$$x[n] = \frac{1}{\sqrt{N}} \sum_{k=-N/2}^{N/2-1} X[k] e^{i(2\pi/N)nk}$$

and the analysis equation or (forward) DFT

$$X[k] = \frac{1}{\sqrt{N}} \sum_{n=0}^{N-1} x[n] e^{i(2\pi/N)nk}$$

The correspondence between the synthesis equation and the equation describing the OFDM symbol should be clear. As long as we restrict the number of subcarriers to less than or equal to N, then we may use an IDFT to synthesize our OFDM waveform. Using the DFT makes the number of samples N in the waveform the same as the number of available subcarriers. If we choose to use less than N subcarriers, then we simply set the remaining inputs to the DFT to zero. Since the FFT implementation of the DFT is so central to practical OFDM systems, such systems are nearly always specified in terms of the FFT or DFT size.

The WiMedia PHY uses a DFT size of 128 points, so that there are 128 subcarriers available. Not all of the subcarriers are used, but we will defer that discussion for a few paragraphs. As mentioned above, the WiMedia PHY sample rate is 528 Msamples/s, so that the subcarrier frequency spacing is $\Delta f = 4.125$ MHz. The symbol time is $T_u = 1/\Delta f$.

Typically, the OFDM waveform is passed through a digital-to-analog (D/A) converter, and then modulated up to the final carrier frequency. The receiver will down-convert back to baseband using the same modulation frequency. If there is no distortion, then the original data may be recovered using a forward DFT. This DFT is simply a bank of matched filters applied to the received signal. Since the individual sinusoids produced by the IDFT produce no mutual interference in the DFT, we have satisfied our orthogonality condition.

A block diagram of the transmit section of the discrete-time portion of our idealized OFDM modem is given in Figure 3.13. The receiver is simply the inverse process. OFDM modems are frequently discussed in terms of *time-domain* and *frequency-domain* processing. Since the modulated symbols are placed on the input to the IDFT, all of the processing before the IDFT has the function of preparing the frequency inputs to the IDFT, and so is considered frequency-domain

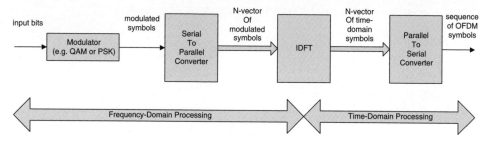

Figure 3.13   Block diagram of the basic OFDM modulator

processing. Similarly, after the DFT, we have time samples and thus time-domain processing.

Obviously, this process is idealized, and we will examine the modems in much more detail in subsequent sections.

3.2.3 Extensions to OFDM

The form of OFDM we have described thus far is the basis for moving forward, but it is seldom used in practice without modification. There are several practical problems that need to be addressed. In this section, we will address some of these issues and show how the OFDM waveform is modified to assist in their solution. In every case, we will reduce the information-carrying capacity of the waveform; we will trade available efficiency in order to simplify the demodulation process.

3.2.3.1 The Zero Postfix and Further ISI Reduction

One modification that is made to most OFDM waveforms comes from recognizing that no matter how far we extend the pulse length, there is still some residual ISI. This is demonstrated in Figure 3.14 for our previous example. The standard OFDM solution is to extend the pulse length by a small amount, but leave the subcarrier frequency spacing alone. As long as the transmit carrier phase is maintained properly, the receiver may discard the extension, and use only the appropriate number of samples for demodulation, thus maintaining orthogonality. Performing the DFT operation as before, but simply repeating the last portion of the waveform at the beginning, may easily accomplish both the pulse-length extension and subcarrier phasing. This approach leads to the term *cyclic prefix* for the pulse extension. As long as the cyclic prefix is longer than the impulse response of the channel, the ISI is removed and the only channel effect that remains is the subcarrier fading. The cyclic prefix process is depicted in Figure 3.15. The cyclic prefix operation is essentially the same as Stockham's overlap-save filtering operation, and operates

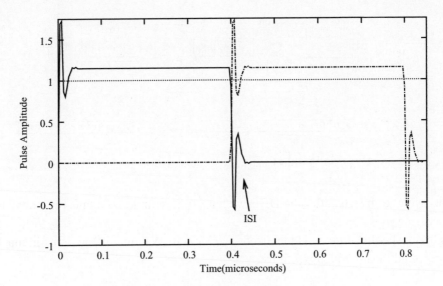

Figure 3.14   Successive pulses demonstrating ISI

by allowing the filter to reach steady state and throwing away the transient period [1]. The cyclic prefix is not used in WiMedia OFDM for the reasons described in the following paragraphs, and details may be found in virtually any paper or book on OFDM systems [2–4].

The cyclic prefix operation lengthens each pulse and, therefore, narrows the spectrum for each subcarrier. The resulting power spectrum, shown in Figure 3.16, has distinct 'scallops' in the frequency response. UWB systems are regulated by

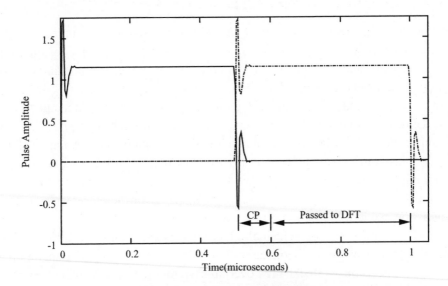

Figure 3.15   Cyclic prefix operation

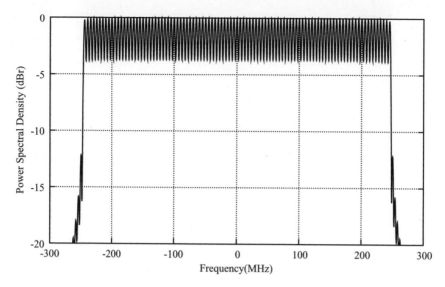

Figure 3.16 Power spectrum of OFDM waveform with cyclic prefix

peak power spectral density, so that any portion of the spectrum that is below the peak results in reduced transmit power. Of course, reduced transmit power results in reduced range and is undesirable.

An alternative to the cyclic prefix is to extend each pulse by adding a short zero postfix (ZP) or zero-padded suffix (ZPS) to the end. Adding zeros to the end of the waveform allows the filtered pulse to settle before starting the next symbol, as shown in Figure 3.17. The transient effects may be removed using the so-called overlap-add technique, in which the signal during the ZP is added to the beginning of the pulse, with the result shown in Figure 3.18. The overlap-add procedure adds the first sample of the ZP to the first sample of the pulse, the second sample of the ZP to the second sample of the pulse, and so forth. As long as the impulse response of the channel is shorter than the ZP, this process recovers the energy lost to the filtering process.

The mathematical expression for the OFDM waveform changes slightly with the addition of the ZP. In particular, the expression for the OFDM subcarrier does not change, but the equation for the transmitted sequence of symbols becomes

$$s[n] = \sum_{m=0}^{M-1} \sum_{k=-K}^{K} d_m[k] e^{i(2\pi/N)k(n-mN_s)} p_N[n - mN_s]$$

where $N_s = N + N_{ZP}$ is the total symbol length, and is the sum of the number of samples in the nonzero OFDM symbol portion of the symbol N and the number of zero samples in the ZP N_{ZP}

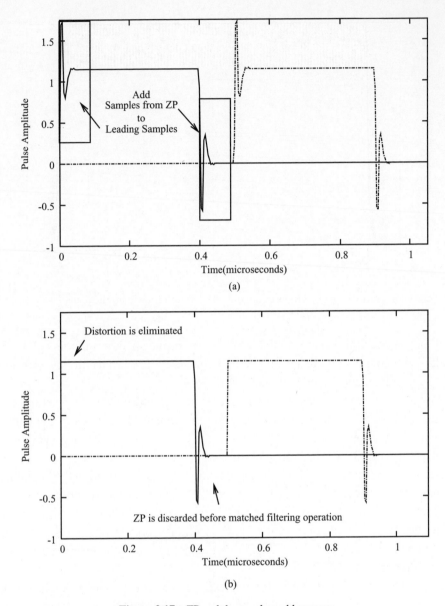

Figure 3.17   ZP and the overlap-add process

In the WiMedia PHY, the number of samples in the OFDM symbol portion of the symbol is 128, as we have already discussed. The length of the ZP is 37 samples, for a total symbol length of 165 samples, or 312.5 ns. Normally, only a portion of the ZP is used for the overlap-add operation, and the remainder to allow time for analog hardware to settle between symbols (i.e. between band hops). The standard does not specify how much of the ZP is to be used for overlap-add, and in fact does

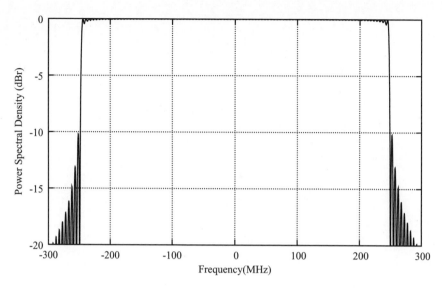

Figure 3.18 Power spectrum of the OFDM waveform with a ZP

not specify the overlap-add procedure at all. The standard does define a constant pBandSwitchTime to be 9.47 ns, or five samples, and states that that the switching time between hops should not exceed pBandSwitchTime. It seems fairly common to use 32 samples for the overlap-add period and leave five samples for settling the PHY layer hardware between frequency hops. Of course, as long as transients are not transmitted and the PPDU is transmitted according to the requirements, it is not possible for an external observer to tell how long the settling time is. On the receive side, the trade becomes a gain in multipath performance versus the cost of settling the hardware quickly, which is a valid implementation trade and again is not easily observed from outside the system.

The WiMedia standard defines the 128-sample output of the FFT to be an OFDM symbol and the 128-sample FFT output plus the 37-sample ZP to be a symbol.

3.2.3.2 Zero, Pilot, and Guard Subcarriers

We have already mentioned that there are 128 subcarriers available in the WiMedia OFDM waveform, but that not all of them are used for data. In addition to data subcarriers, some of the subcarriers are zero, some are used as *pilot* subcarriers, and others are used as *guard* subcarriers. In this section, we define these terms and discuss the structure and rationale for these elements.

Figure 3.19 shows the structure of the subcarriers for the higher subcarrier frequencies. Note that the figure, and in many cases the WiMedia PHY specification,

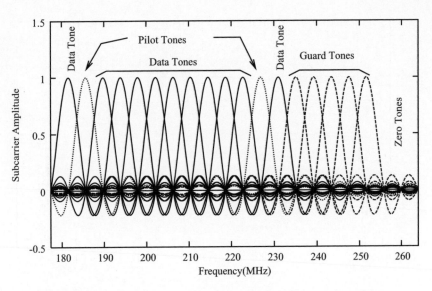

Figure 3.19 High-frequency subcarriers showing pilot, guards, and data

uses the terms *tone* and *subcarrier* interchangeably simply because the subcarrier is a pulse at a constant frequency for the duration of the OFDM symbol.

The analog sections of most radios, due to mismatches, local oscillator feed-through, and other imperfections, induce a low-frequency component, usually called a *DC component*, into the received waveform. This DC component makes detection of the zero-frequency subcarrier quite difficult, and so the zero-frequency term is usually left out; that is, we set $X[0] = 0$.

Another serious issue with many OFDM systems, and the WiMedia system in particular, is a result of two OFDM signals that are side by side in frequency; that is, signals that occupy *adjacent bands*. We have so far only briefly discussed the frequency-hopping nature of the WiMedia waveform. The center frequencies of adjacent frequency bands are spaced exactly 528 MHz apart. Since the subcarriers are 4.125 MHz apart, then if we use all 128 of them we see that there is no space between the adjacent bands. Without some space, it will not be possible to filter the adjacent band out before sampling at 528 MHz, and the adjacent band will alias into the desired signal.

To deal with aliasing of the adjacent band, the outer subcarriers are set to zero. In particular, we use $K = 61$ in our equation for the OFDM waveform. Since the DC term is also set to zero, this leaves 122 active subcarriers in the OFDM waveform.

Even setting the outer subcarriers to zero leaves only five zero-subcarriers, or just over 20 MHz between adjacent bands. This is certainly not enough for an analog filter to reduce the adjacent band sufficiently to avoid aliasing. Many implementations to date use a higher analog-to-digital (A/D) sampling rate (e.g. 1056

Msamples/s) and then use a digital filter and decimator to reduce the sample rate to 528 Msamples/s.

Even with a digital filter, a 20 MHz dead zone between adjacent bands calls for fairly aggressive filtering, but adding a larger dead zone will reduce the signal bandwidth to less than the 500 MHz required by the FCC and other regulatory agencies. To simplify the adjacent-band filtering problem, the WiMedia OFDM waveform allocates five subcarriers on either end of the spectrum to be *guard* sub-carriers. In other words, $X[k]$, for $k = \{-61, -60, -59, -58, -57, 57, 58, 59, 60, 61\}$, are guard carriers. In early versions of the standard, these were simply predefined values for spectrum-shaping purposes. However, the FCC and other regulatory agencies rightfully frown on the idea of taking up spectrum just for the sake of adding bandwidth, so WiMedia chose the simple artifice of duplicating data subcarriers, as shown in Figure 3.20. The implementer may reduce the power on the guard subcarriers, for example, to shape the spectrum to simplify the filtering problem described above. Of course, it is the implementer's responsibility to ensure that the resulting waveform satisfies regulatory requirements. The standard does not impose any requirement that the receiver should use these subcarriers in any way. Many implementations filter the guards, thus relaxing the filtering requirements on the receiver.

The final modification we will make to the basic OFDM system is to add *pilot* subcarriers. Pilot subcarriers are simply modulated subcarriers in which the receiver knows the modulated symbols a priori. These modulated symbols provide

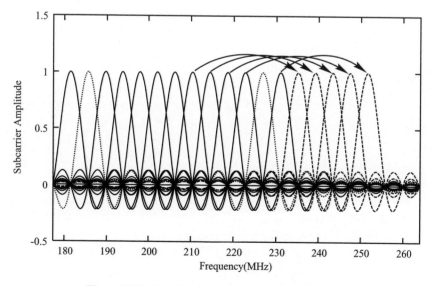

Figure 3.20   Copying the data tones to the guard tones

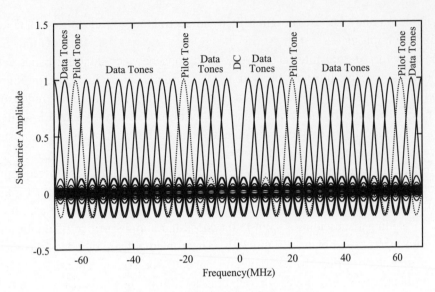

Figure 3.21 WiMedia subcarrier structure

a reference by which the receiver may maintain estimates of signal parameters without the additional uncertainty of estimating the modulated data as well.

In many OFDM systems, pilots are spaced as far apart as possible, but close enough together that the receiver is able to infer some tracking information, e.g. via interpolation, about nearby subcarriers from the pilot tones. In WiMedia OFDM, the pilots are placed every ten subcarriers or 41.25 MHz apart. In many UWB channels, the correlation between the pilots and the nearby carriers is reasonably strong. In the more severe channels the correlation is fairly weak.

The pilots are placed at subcarrier numbers $\{-55, -45, -35, -25, -15, -5,$ $5, 15, 25, 35, 45, 55\}$. We will not repeat the details of the standard here, but the reader should note that the pilots are designed to be conjugate symmetric in frequency. The importance of this will become clear later. Furthermore, the pilots are not constant from symbol to symbol, but are *scrambled* to reduce spectral spikes and other spectral artifacts. Figure 3.21 depicts the pilot tones for the lower frequency subcarriers.

3.2.4 Spreading

The WiMedia PHY standard introduces the concept of *spreading* into the OFDM waveform. Simply put, spreading is a technique in which the same data is placed on multiple subcarriers. Spreading is introduced as a technique for providing diversity in the face of frequency-selective fading.

The WiMedia standard has two types of spreading: time spreading and frequency spreading. There are three modes in the WiMedia standard: (1) time and frequency spreading used together; (2) time spreading with no frequency spreading; and (3) no spreading at all. That is, when frequency spreading is used, time spreading is always used as well, but time spreading may be used without frequency spreading. To complicate matters further, when time spreading is used without frequency spreading, the mechanism is slightly different than when both techniques are used together.

Frequency spreading consists of placing the same data symbol on frequency-symmetric subcarriers within the same OFDM symbol. In other words, if a data symbol is placed on the kth subcarrier, then the same data symbol will be conjugated and then placed on the $-k$th subcarrier as well. Since the pilots are already conjugate symmetric, this ensures that the waveform will be conjugate symmetric when frequency spreading is used. A conjugate symmetric signal has a zero imaginary part, allowing the D/A converter for the quadrature channel to be turned off on the transmitter when frequency spreading is used. The idea behind frequency spreading is to reduce the probability that both subcarriers will be impacted by amplitude fading. Of course, regardless of the specifics, both subcarriers will have a different fading level; thus, a recombining algorithm (e.g. maximal-ratio combining) could be used by the receiver.

When frequency spreading is used, time spreading simply duplicates the symbol in time. In other words, the $2m$th symbol is simply copied into the $(2m + 1)$th symbol, but with a twist. The data symbols that form the first symbol are used to form the second symbol as well. However, the data symbols are scrambled before creating the second symbol to reduce correlation between adjacent symbols, thereby reducing the spectral artifacts that are present in the presence of correlation. In the case where frequency hopping is present, this technique provides four-way diversity for each data symbol, since each data symbol is on four distinct transmit frequencies. The WiMedia PHY standard has modes in which the hopping is turned off. In this case, the technique only provides two-way diversity, but still doubles the energy per bit and thus enhances link performance. Figure 3.22 shows the mapping of $X[4]$ into subcarriers $k = 4$ and $k = -4$ for two consecutive OFDM symbols.

When time spreading is used without frequency spreading, the data symbol on the kth subcarrier of the $2m$th symbol is duplicated onto the $-k$th subcarrier of the $(2m + 1)$st symbol. The duplication is performed by swapping the real and imaginary parts and scrambling the result. Again, this technique reduces correlation between adjacent OFDM symbols. Note that, in this case, even without frequency hopping, time spreading provides two-way diversity. Figure 3.23 shows the

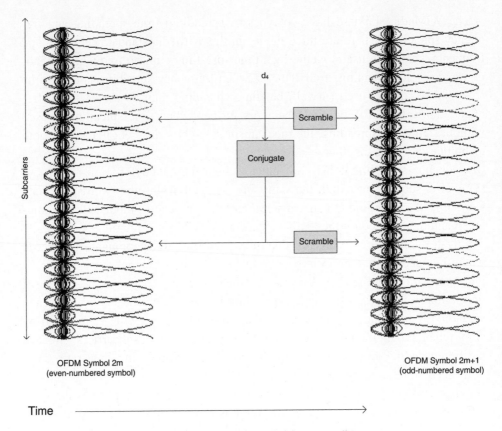

Figure 3.22   Frequency- and time-spreading

mapping of $X[4]$ into subcarriers $k = 4$ and $k = -4$ for two consecutive OFDM symbols.

3.2.5 Modulation

The modulated data symbols $X_m[k]$ are complex numbers that are functions of the bits that are received from the outer modem and are created by using one of two modulation techniques: QPSK and a new technique called DCM. Recall that the outer sublayer adds error correction and does other bit-level processing on the data received from the MAC sublayer. The inner sublayer breaks the coded bits into groups of N_c bits and maps each group of bits into an OFDM symbol. The value of N_c depends on the modulation type and the amount of spreading. The WiMedia inner sublayer has three basic modes of operation, with specific modulation types and spreading in each mode as shown in Table 3.1. We will label the coded bits

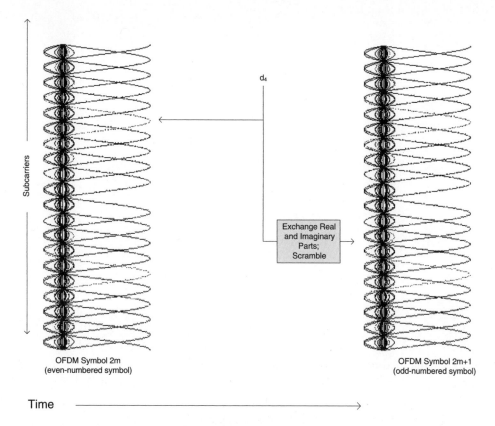

OFDM Symbol 2m
(even-numbered symbol)

OFDM Symbol 2m+1
(odd-numbered symbol)

Exchange Real
and Imaginary
Parts;
Scramble

d_4

Subcarriers

Time

Figure 3.23 Time spreading

$b[n]$, $0 \leq n < N_{\mathrm{c}}$, with each bit taking on the values 0 and 1. Before modulating, each bit is converted to ± 1 using the formula $\beta[n] = 2b[n] - 1$.

The simplest form of modulation used in the WiMedia PHY is textbook QPSK. Each pair of bits is mapped to a complex number using the standard Gray-coded approach shown in Figure 3.24. The resulting complex number is normalized by $1/\sqrt{2}$ so that each symbol has unit energy. This process may be written as

$$X[k] = \frac{1}{\sqrt{2}} (\beta[2n] + i\beta[2n + 1])$$

Table 3.1 Inner sublayer modes of operation for even-numbered symbols

Mode	Modulation	Spreading	N_{c}	Conjugate symmetric?
Mode 1	QPSK	Time and frequency	100	Yes
Mode 2	QPSK	Time	200	No
Mode 3	DCM	None	200	No

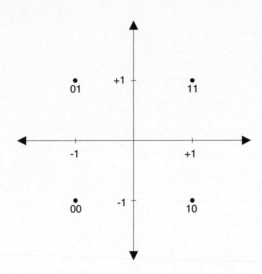

Figure 3.24   QPSK constellation and bit mapping

The value of k is precisely described in the WiMedia PHY standard, but is easily understood without the complex mathematics. The first QPSK symbol, i.e. for $n = 0$, is mapped into the lowest frequency OFDM *data* subcarrier, or subcarrier number -56. The second QPSK symbol is mapped into the next OFDM data subcarrier, subcarrier -54 (remember that subcarrier -55 is a pilot and does not carry data). The process continues exactly like this until all the QPSK symbols are used. Of course, in Mode 1, we use frequency spreading and so the first QPSK symbol is conjugated and mapped into subcarrier 56 as well, etc.

Both of the QPSK modes Mode 1 and Mode 2 use time spreading. With time spreading, the same information is duplicated from the even symbol to the next odd symbol. This means that the number of code bits per two symbols is the same as the number of code bits for one symbol. As we will see a little later, it is more natural to group symbols in terms of six symbols rather than two, and we adopt that convention here. To simplify Table 3.1, we define $\bar{N}_c$ to be the number of coded bits per six OFDM symbols, resulting in Table 3.2.

Table 3.2   Inner sublayer modes of operation per group of six OFDM symbols

Mode	Modulation	Spreading	$\bar{N}_c$	Conjugate symmetric?	Implementation mandatory?
Mode 1	QPSK	Time and frequency	300	Yes	Yes
Mode 2	QPSK	Time	600	No	Yes
Mode 3	DCM	None	1200	No	No

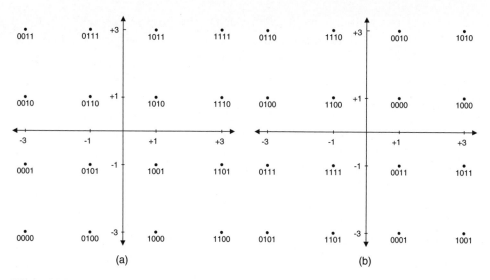

Figure 3.25 Dual-carrier modulation constellation and bit mapping: (a) first subcarrier mapping; (b) Second subcarrier mapping

The PHY standard introduces a new form of modulation called DCM, which takes four bits and creates two $X[k]$ values that are placed on a pair of subcarriers. Figure 3.25 depicts the mapping for each possible combination of four bits. Figure 3.25a shows the mapping for the first generated subcarrier and Figure 3.25b shows the mapping for the second subcarrier. Implementation of DCM is optional.

The DCM symbols for subcarriers k_1 and k_2 may be expressed as

$$X[k_1] = \frac{1}{\sqrt{10}}(2\beta[n_1] + \beta[n_1 + 1]) + i(2\beta[n_2] + \beta[n_2 + 1])$$

and

$$X[k_2] = \frac{1}{\sqrt{10}}(\beta[n_1] - 2\beta[n_1 + 1]) + i(\beta[n_2] - 2\beta[n_2 + 1])$$

The values for n_1 and n_2 are selected from a complicated-looking formula. However, they can be simply explained. To begin with, n_1 is set to 0 and n_2 is set to 50, and the first symbol pair is formed. Then, both n_1 and n_2 are incremented by 2 and the second symbol is formed. When n_1 reaches 50, we reset n_1 to 100 and n_2 to 150 and continue in the same fashion until all of the bits are used. The values for k_1 and k_2 are derived much like those for QPSK. For the first symbol ($n_1 = 0$), we map the first symbol to the lowest frequency subcarrier, or $k_1 = -56$, and set k_2 to the lowest positive frequency subcarrier, or $k_2 = 1$. The next symbol pair is mapped into the next available data subcarrier, or $k_1 = -54$ and $k_2 = 2$. The process continues, skipping pilots as needed, until all of the bits are used and all of the data subcarriers are filled.

It is easily demonstrated that DCM is equivalent to QPSK in a flat AWGN channel, i.e. an AWGN channel in which all of the received subcarriers have the same amplitude. However, in an environment in which the subcarriers have different levels of fading, DCM provides a level of diversity similar to frequency spreading but with greater spectral efficiency.

3.2.6 Tone Nulling

UWB devices take an enormous amount of spectrum and, in spite of the very low power levels, there is a concern (especially among regulators) that they may interfere with other (narrower band) devices that occupy the same spectrum. Furthermore, other devices that occupy the UWB spectrum may interfere with the UWB waveform. Since OFDM is formed by modulating individual subcarriers, we have the possibility of dropping subcarriers that occupy the same spectrum as narrowband devices. The WiMedia specification includes an optional mechanism for zeroing tones as a means of avoiding certain frequency ranges.

The PHY nulls tones according to a mask provided by the upper network layers. The mask is simply an array of bits, one for each tone in the band group. The first 128 bits are for the lowest frequency band in the band group, the second 128 bits for the middle band, and the final 128 bits for the upper band (if there is an upper band in the band group).

The PHY will perform all of the data translation up to the modulation, pilot insertion, etc., and simply apply the mask. The specification does not say that the tones are zeroed, or imply any implementation, nor does it specify any power reduction in the nulled subcarriers. It is up to the implementer to see to it that the result meets any regulatory requirements. The reader should realize that zeroing a tone does not imply zero transmitted energy in that frequency band, since adjacent tones 'overflow' into the band. See Figure 3.21, in which the zero-frequency subcarrier is nulled and the sidelobes from adjacent tones are clearly present.

Since tone nulling is performed after the coding, modulation, spreading, etc., it is possible that data symbols are lost. The only recovery mechanism at the PHY layer is error correction. Of course, it is possible to use erasures in the decoder, and the specification requires that 86 useful tones be generated, regardless, in any band. Furthermore, spreading may duplicate the tone on another subcarrier.

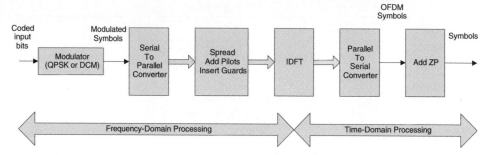

Figure 3.26   WiMedia OFDM modulator

The specification also allows the PHY to null other tones to maintain conjugate symmetry or for other purposes.

3.2.7 Summary

We have now completely specified the form of the OFDM modulator, or inner sub-layer. A block diagram is shown in Figure 3.26. The WiMedia standard introduces some unique features to the OFDM technique, including DCM and a ZP instead of a cyclic prefix. The OFDM parameters are shown in Tables 3.3 and 3.4. Here, the numbers in Table 3.4 may be derived from those in Table 3.3 using the formulas shown in the Table 3.4.

3.3 PMD

As explained before, the WiMedia PHY specification does not explicitly define where the PLCP sublayer ends and where the PMD sublayer begins. However,

Table 3.3   Core OFDM parameters

Parameter	Value	Description
f_s	528 MHz	Sampling frequency or system bandwidth
N	128	FFT size or number of available subcarriers or number of samples in OFDM symbol (FFT output)
N_G	10	Number of guard subcarriers, subcarrier numbers $\{-61,-60,-59,-58,-57,57,58,59,60,61\}$
N_P	12	Number of pilot subcarriers, subcarrier numbers $\{-55,-45,-35,-25,-15,-5,5,15,25,35,45,55\}$
N_Z	6	Number of zero subcarriers, subcarrier numbers $\{-64,-63,-62,0,62,63\}$
N_{ZP}	37	Number of samples in zero postfix
N_{BS}	5	Maximum number of samples allowed for band switch

Table 3.4 Derived OFDM parameters

Parameter	Value	Description
N_D	100	Number of data-bearing subcarriers $N_D = N - N_P - N_G - N_Z$
N_T	122	Total number of used subcarriers $N_T = N_P + N_G + N_D$
Δt	1.89 ns	Sample interval $\Delta t = 1/f_s$
Δf	4.125 MHz	Subcarrier spacing $\Delta f = f_s/N$
N_s	165	Number of samples per symbol $N_s = N + N_{ZP}$
T_s	312.5 ns	Total symbol time $T_s = N_s \Delta t$

the specification does state that a number of components are in the PLCP, so we will simply assume that everything not specifically included in the PLCP is in the PMD. Thus, PMD consists of:

- data converters (A/D and D/A);
- frequency hopping, including mixers and synthesizers;
- amplifiers;
- automatic gain control (AGC);
- antenna;
- filters.

The specification does not, of course, prescribe any specifics for the implementation, but describes the interface at the antenna.

3.3.1 WiMedia Band Plan

The WiMedia PHY specification divides the UWB spectrum into 14 *bands*. Each band is 528 MHz wide and centered on odd multiples of 264 MHz. The 528 MHz spacing ensures that a WiMedia OFDM symbol fits exactly in a band. The center frequencies were chosen to maximize the number of bands that fit into the UWB spectrum. The center frequency for band k is $f_k = (2k + 11)264$ MHz.

ECMA-368 limits the operation of the PHY layer to one of six *band groups*, each consisting of two or three bands. Band Group 1 consists of Band 1, Band 2, and Band 3, and Band Group 2 consists of Band 4, Band 5, and Band 6, etc., as shown in Figure 3.27 and Table 3.5. Band Group 5 has only two bands. Band Group 6 overlaps Band Groups 3 and 4 and was created to mesh with certain worldwide regulatory requirements (see Chapter 2).

3.3.2 Frequency Hopping

The WiMedia-compliant PHY must have the ability to frequency hop. The hopping only occurs within the bands in a single band group. Thus, for most of the band groups, the instantaneous PHY RF bandwidth is 528 MHz and the total active RF

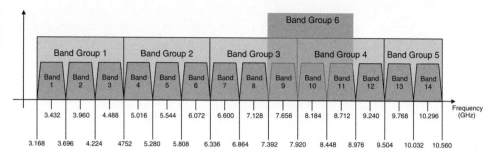

Figure 3.27 WiMedia bands and band groups

bandwidth is three times that, or 1584 MHz. For Band Group 5, there are only two bands, so the total active RF bandwidth is 1056 MHz. Frequency hopping the OFDM signal adds considerable complexity to the RF frontend, but does mean that the data converters need not be designed to sample the entire 1584 MHz available bandwidth. Worldwide regulations insist that UWB signals have limited power spectral density, but this is measured over relatively long time-frames (relative to the UWB symbol times). Thus, frequency hopping over three bands allows the fundamental PHY sampling rate to remain 528 MHz, but allows for three times the instantaneous transmit power.

The WiMedia specification describes 10 basic hopping patterns, termed TFCs, shown in Table 3.6. For example, when using TFC 1, the first OFDM symbol will be transmitted on center frequency $f_1 = 3432$ MHz, the second symbol

Table 3.5 WiMedia bands and band groups

Band group	Band	Center frequency (MHz)
1	1	3432
	2	3960
	3	4488
2	4	5016
	5	5544
	6	6072
3	7	6600
	8	7128
	9	7656
4	10	8184
	11	8712
	12	9240
5	13	9768
	14	10296
6	9	7656
	10	8184
	11	8712

Table 3.6 TFCs

TFC	Hopping pattern					
1	1	2	3	1	2	3
2	1	3	2	1	3	2
3	1	1	2	2	3	3
4	1	1	3	3	2	2
5	1	1	1	1	1	1
6	2	2	2	2	2	2
7	3	3	3	3	3	3
8	1	2	1	2	1	2
9	1	3	1	3	1	3
10	2	3	2	3	2	3

$f_2 = 3960$ MHz and the third on $f_3 = 4488$ MHz, then the pattern will repeat for the entire packet. Note that TFCs 1–4 and 8–10 are the actual hopping patterns, but TFCs 5–7 are fixed-frequency transmissions. TFCs 1–4 are Time–Frequency Interleaved (TFI) modes, and allow for full power transmission as described above. TFCs 8–10 are also TFI modes, but hop over only two frequencies and so are termed TFI2 modes, and must reduce power to two-thirds of full power. TFCs 5–7 are fixed-frequency patterns, called FFI modes and must reduce power to one-third full power.

The discussion in the previous paragraph assumes that transmission is occurring in Band Group 1, but to transmit in a different band group, the patterns are simply shifted up to the center frequencies for that band group. Thus, TFC 1 in Band Group 3 hops in the repeated pattern of $f_7 = 6600$ MHz, $f_8 = 7128$ MHz, and $f_9 = 7656$ MHz or Bands 4, 5, and 6. Band Group 5 has only two bands, and so is restricted to TFCs 5–8.

The hopping occurs at the end of each OFDM symbol, sometime during the ZP. The standard requires that the band switch time be less than or equal to five samples, or 9.47 ns. All of the hopping patterns have a common period of six symbols, which we will term a *hop frame*. All PHY transmissions are multiples of six symbols, and thus all relevant specifications are in terms of six symbols.

Finally, the PHY specification defines a group of *channels*. A PHY transmitting in channel number 9 (channels 1–8 are not defined and reserved for future use) is transmitting in Band Group 1, using TFC 1. Channel 10 is Band Group 1, TFC 2, channel 17 is Band Group 2, TFC 1, etc. The reader is referred to the ECMA-368 specification for details. The channel number uniquely identifies both the band group and the TFC. Typically in communication systems, when two radios are transmitting in two different channels, there is minimal interference between the two. The rationale behind the WiMedia PHY channel specifications is the same: two PHYs transmitting on separate channels are nominally independent. Of course, with hop sequences of only three center frequencies, two PHYs transmitting in the same band group will frequently collide one-third of the time, so the concept of channel here is somewhat dubious.

As a final note on frequency hopping and center frequencies, the WiMedia standard specifies that the transmitted RF signal satisfy

$$s_{RF}(t) = \sum s(t - nT_s)e^{i2\pi f_c(q(n))t}$$
$$= \sum s_I(t - nT_s)\cos[2\pi f_c(q(n))t] - is_Q(t - nT_s)\sin[2\pi f_c(q(n))t]$$

where all the elements of this equation have been discussed previously. Note the sign of the imaginary part is negative, and care must be taken to ensure that it remains negative when converting the baseband signal to RF and back to baseband.

3.3.3 Common Clock Reference

Like most OFDM specifications, the WiMedia PHY specification expressly states that frequencies and clocks for the PHY layer must be derived from the same reference. That means that center frequencies and sampling clock must be derived from the same reference, usually a crystal oscillator. Since all of the frequencies are multiples of 33 MHz, many implementers are using 33 MHz crystals for the reference, but there is no requirement in the specification for this. The center frequencies and sampling clock must be within 20 ppm (parts per million) of the specification.

An obscure but important feature is that the phase of the transmitted carriers be coherent from hop to hop within a packet. Figure 3.28 shows the meaning of this clause. Whenever the center frequency hops away from a given frequency, the phase must be the same when it returns to the starting frequency. This is simply to avoid the necessity for recomputing phase estimates on each hop.

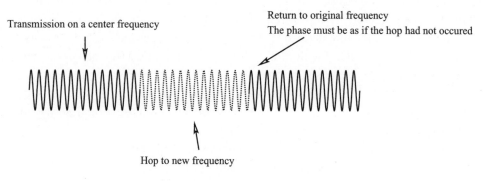

Figure 3.28   Hop-to-hop coherence

Table 3.7   Transmit power levels

Level	Attenuation for TFI modes (dB)	Attenuation for TFI2 modes (dB)	Attenuation for FFI modes (dB)
0	0	0	0
1	2	2	2
2	4	4	4
3	5	5	6
4	8	8	8
5	10	10	Reserved
6	12	Reserved	Reserved
7	Reserved	Reserved	Reserved

3.3.4 Transmit Power Control

The WiMedia PHY specification suggests (but does not require) that the PHY layer implement transmit power control. Table 3.7 shows the levels. The levels are required to be monotonic and have an accuracy of ± 20 %, with a worst-case accuracy of ± 1 dB. The number of levels required for FFI modes is less than for TFI modes since the PHY is already required by regulatory agencies to reduce the power off by one-third (4.8 dB). Similarly, TFI2 modes must reduce power relative to the TFI modes by two-thirds (1.8 dB).

3.4 Scrambling, Interleaving, and Error Correction

In Section 3.2.1 we showed that OFDM reduces the frequency-selective fading problem to a set of flat fading ones. It can further be shown that the flat fading in a UWB channel may be approximated by Rayleigh fading: the amplitude of each subcarrier is distributed approximately as a Rayleigh random variable. Reliable communication in a Rayleigh fading channel requires the use of error correction [5]. ECMA-368 (PHY and MAC) uses two forms of error correction. The MAC sublayer provides feedback error correction at the packet level by retransmitting failed packets (automatic repeat-request (ARQ)) and the PHY provides FEC using both convolutional and Reed–Solomon coding so that the overall strategy is a hybrid ARQ approach. In addition to error correction, ECMA-368 uses a frame check sequence (FCS) – usually computed by MAC – to detect frame errors. In this section, we discuss the PHY error correction and related bit-level processing. Using the terminology described earlier in this chapter, this section describes the elements that comprise the PHY outer sublayer.

The outer sublayer consists of (a) scrambling the input bits, (b) adding an FCS, (c) adding tail and pad bits, and (d) scrambling the result. In the rest of this section we will discuss the algorithmic structure of these elements. The processing discussed here is not applied to all of the packet elements, but we will discuss the specifics in Section 3.5.

3.4.1 Scrambling

As we have mentioned before, in our discussion of pilots and spreading, the presence of correlation in the modulated symbols placed on the OFDM subcarriers may induce spectral artifacts, such as spikes and nulls. This is almost always undesirable, but in the UWB channel such spikes can be a serious problem. The transmit power of a UWB signal is determined by the peak power spectral density, so spikes may force the designer to reduce power simply to manage spikes.

The use of scrambling is a simple and effective method to reduce spectral artifacts. We have already seen the use of scrambling in the placement of pilot tones and in the time-spreading process. In order to minimize correlation, we further scramble the input information bits before any other processing. Figure 3.29 shows the scrambling process. The input data is exclusive-OR-ed with a long pseudorandom bit stream (PRBS). Since the PRBS has minimal correlation, the resulting scrambled data will also have minimal correlation. Figure 3.29 also shows that scrambling is its own inverse. Simply reapplying the scrambling sequence at the receiver end will recover the data. Of course, Figure 3.29 does not show the intervening error correction and interleaving, nor does it show the OFDM and channel process.

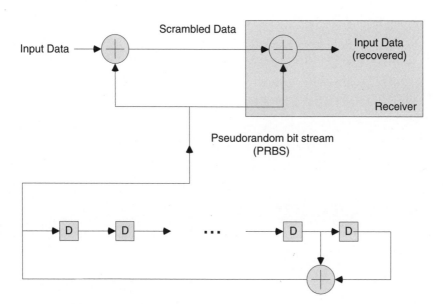

Figure 3.29 Data scrambler

The PRBS generator is simply a long maximal-sequence generator consisting of a 15-element shift register and two feedback taps. The initial content or *seed* of the shift register makes a significant difference on the resulting scrambled sequence, and the WiMedia standard provides for four different seed values. The standard further specifies that consecutive packets must be sent with different seeds. One reason for this is that, in addition to random noise errors, there are other forms of error, such as clipping and other transceiver-induced errors. Using the same seed for a retransmitted packet could reproduce the source of the error, causing the error to reoccur. Of course, the receiver must use the same seed value as the transmitter for the descrambling to work properly.

3.4.2 Convolutional Coding

FEC coding in the WiMedia PHY is a straightforward but powerful convolutional code that is punctured to achieve several code rates. The base is a rate 1/3 code that has been used in many applications, including some NASA deep-space missions [5]. This rate 1/3 code is punctured to rates 1/2, 5/8, and 3/4. In Table 3.2 we described the modes of operation for the inner sublayer. Table 3.8 shows how the code rates are applied to the different inner sublayer modes. Recall that we defined $\bar{N}_c$ to be the number of coded bits per hop frame, or per six OFDM symbols. We introduce the *spreading factors* S_T and S_F as the amounts of time and frequency spreading. Each of these terms may be either 1 or 2, and defines the number of subcarriers assigned to a modulated symbol. We now define $\bar{N}_b$ to be the number of information bits per hop frame. Of course, $\bar{N}_b = R_c \bar{N}_c$, where R_c is the code rate. Since each symbol is $T_s = 312.5$ ns long, we may now define the data rates offered by WiMedia PHY as $R_d = \bar{N}_b/6T_s$, also displayed in Table 3.8. Finally, we attach an index, which we will call the Rate ID or rate index, to each of the data rates. Since implementation of DCM is optional, the highest rates (Mode 3 rates) are optional.

Table 3.8 WiMedia data-rate table

Mode	Modulation	Spreading	$\bar{N}_c$	R_c	$\bar{N}_b$	R_d (Mbits/s)	Rate ID
Mode 1	QPSK	$S_T = 2$ and	300	1/3, 1/2	100	53.3	0
		$S_F = 2$			150	80	1
Mode 2	QPSK	$S_T = 2$ and	600	1/3, 1/2, 5/8	200	106.7	2
		$S_F = 1$			300	160	3
					375	200	4
Mode 3	DCM	$S_T = 1$ and	1200	1/2, 5/8, 3/4	600	320	5
		$S_F = 1$			750	400	6
					900	480	7

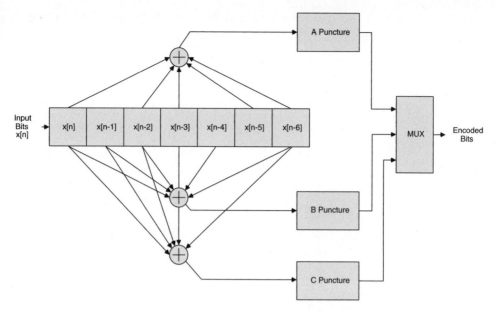

Figure 3.30 Convolutional code

The convolutional code is based on the code polynomials with octave values 133, 165, and 171 or binary patterns 1011011, 1110101, and 1111001. The binary patterns determine the weights in a feed-forward shift register, as shown in Figure 3.30. The initial state of the shift register is always set to zeros at the start of encoding. The rate 1/3 code was selected because rate $1/N$ codes are simple to implement, and a rate 1/3 code is powerful enough to create a system with excellent performance in fading channels and some protection against multiuser interference.

Puncturing is accomplished by deleting certain bits from the coded bit streams. The puncture patterns are shown in Table 3.9, with a 1 representing a bit that is kept and a 0 representing a bit that is deleted. Each input data bit produces three code bits, and the puncturing works by applying the puncture pattern in sequence to each of the three code bits. For the first set of three code bits, the first column of the puncture sequence is applied and deleted bits are not multiplexed into the

Table 3.9 Puncture patterns

Code rate	1/2	5/8	3/4
A puncture pattern	1	10101	100
B puncture pattern	0	10101	100
C puncture pattern	1	01010	011

output bit stream. The second set of code bits will have the second column of the puncture sequence applied and, when all columns have been used, the process will start with the first column. Note that the rate 1/2 code deletes the entire 'B' code bit stream, so that the rate 1/2 code is simply a convolutional code with polynomials 133 and 171. This is another standard rate $1/N$ code [5].

The receiver will depuncture by placing dummy bits in the matched filter stream to the decoder; for example, by inserting zeros for the matched filter input.

Let us examine an example of the encoding, puncturing, and depuncturing process. Consider a set of input bits

$$b = \{1, 1, 1, 0, 1, 0, 1, 1, 0, 0\}$$

The output of the encoder is

$$C = \begin{bmatrix} 1 & 1 & 0 & 0 & 1 & 0 & 0 & 0 & 0 & 0 \\ 1 & 0 & 1 & 0 & 1 & 0 & 0 & 1 & 0 & 1 \\ 1 & 0 & 1 & 1 & 1 & 0 & 1 & 0 & 1 & 0 \end{bmatrix}$$

where the top row is the output of the first (A) polynomial, the middle row is the output of the second (B) polynomial, and the bottom row is the output of the third (C) polynomial. After multiplexing, the encoded (rate 1/3) bit stream is

$$c = \{1, 1, 1, 1, 0, 0, 0, 1, 1, 0, 0, 1, 1, 1, 1, 0, 0, 0, 0, 0, 1, 0, 1, 0, 0, 0, 1, 0, 1, 0\}$$

Applying the rate 1/2 puncturing pattern, we have

$$C = \begin{bmatrix} 1 & 1 & 0 & 0 & 1 & 0 & 0 & 0 & 0 & 0 \\ x & x & x & x & x & x & x & x & x & x \\ 1 & 0 & 1 & 1 & 1 & 0 & 1 & 0 & 1 & 0 \end{bmatrix}$$

where the x entries indicate a deleted bit. After multiplexing, the encoded (rate 1/2) bit stream is

$$c = \{1, 1, 1, 0, 0, 1, 0, 1, 1, 1, 0, 0, 0, 1, 0, 0, 0, 1, 1, 0\}$$

The depuncturing process yields the same matrix as above, except that the x entries are inserted dummy bits.

The result of applying the rate 5/8 puncturing pattern is

$$C = \begin{bmatrix} 1 & x & 0 & x & 1 & 0 & x & 0 & x & 0 \\ 1 & x & 1 & x & 1 & 0 & x & 1 & x & 1 \\ x & 0 & x & 1 & x & x & 1 & x & 1 & x \end{bmatrix}$$

or, after multiplexing

$$c = \{1, 1, 0, 0, 1, 1, 1, 1, 0, 0, 1, 0, 1, 1, 1, 1\}$$

Finally, the rate 3/4 puncturing pattern yields

$$C = \begin{bmatrix} 1 & x & x & 0 & x & x & 0 & x & x & 0 \\ 1 & x & x & 0 & x & x & 0 & x & x & 1 \\ x & 0 & 1 & x & 1 & 0 & x & 0 & 1 & x \end{bmatrix}$$

or

$$c = \{1, 1, 0, 1, 0, 0, 1, 0, 0, 0, 0, 1, 1, 1\}$$

A small but important aspect of the convolutional code is the use of tail bits. Tail bits are simply zero information bits added at the end of the information sequence that force the convolutional encoder to return to the zero (or initial) state. This simply allows the decoder to have a known state with which to begin the traceback function, reducing some decoder uncertainty and, therefore, adding to the coding performance (coding gain). Since the shift register is 6 bits long, it takes six tail bits to drive the encoder to the zero state.

3.4.3 Header Protection: Header Check Sequence and Reed–Solomon Coding

The payload is protected from errors by an FCS, with error correction provided by the convolutional code described above. Since the header contains data rate information, scrambling seeds, etc., even a minor error in the header almost certainly means loss of the entire packet. Consequently, the header has a check sequence of its own and, in addition to the convolutional code, has a concatenated Reed–Solomon code.

The header check sequence (HCS) is a simple two-octet cyclic redundancy check sequence and is depicted in Figure 3.31.

The Reed–Solomon code is a fairly standard systematic (255, 249) code over $GF(2^8)$, or a code in which the data elements and the parity elements are 8-bit quantities (octets). A Reed–Solomon code is a block code, in this case taking a block of 249 octets and producing a block of 255 octets. Since the code is systematic, the first 249 octets are simply the message and the remainder are parity octets. Because the header is much shorter than 249 octets, the message is prepended with enough zeros to fill out the 249 octets, resulting in a shortened Reed–Solomon code.

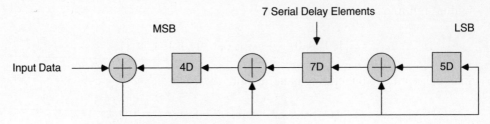

1. Preset 16-element shift register to all ones
2. Input the header data
3. Read the shift register, MSB first
4. Complement the result to form the HCS

Figure 3.31 Header check sequence

The encoding process (Figure 3.32) is performed by computing the remainder of the message polynomial divided by the generator polynomial

$$g(x) = x^6 + g_5 x^5 + g_4 x^4 + g_3 x^3 + g_2 x^2 + g_1 x^1 + g_0$$

with the values of g_k given in Table 3.10. The operations are all performed in GF(2^8) [5].

For both the Reed–Solomon code and the HCS, sample computations are found in Annex E of the ECMA-368 standard. Although the generation of Reed–Solomon code in the transmitter is mandatory, there is no requirement for the receiver to use it for the purpose of error correction.

3.4.4 Interleaving

Convolutional codes provide maximum protection against random errors, or errors that are distributed uniformly throughout the frame. They do not provide good protection against errors that come in bursts, and interleaving is used to convert burst errors to random errors.

Table 3.10 Reed–Solomon generator polynomial coefficients

Coefficient	Value (decimal)
g_0	117
g_1	49
g_2	58
g_3	158
g_4	4
g_5	126

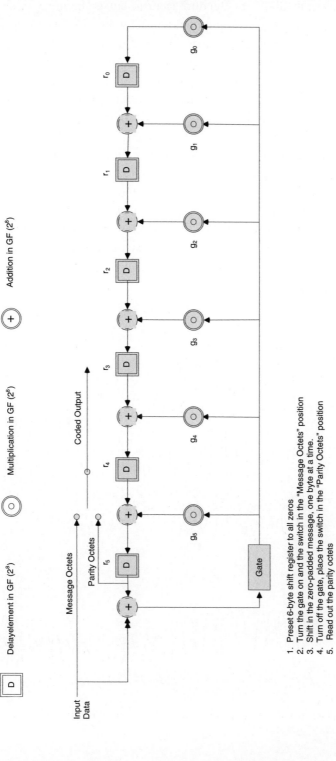

Delayelement in GF (2^8)

Multiplication in GF (2^8)

Addition in GF (2^8)

1. Preset 6-byte shift register to all zeros
2. Turn the gate on and the switch in the "Message Octets" position
3. Shift in the zero-padded message, one byte at a time.
4. Turn off the gate, place the switch in the "Parity Octets" position
5. Read out the parity octets

Figure 3.32 Reed–Solomon encoder

In most wireless media, errors tend to occur in bursts. Figure 3.33 shows portions of the frequency-domain representation of two OFDM symbols. The first symbol shows fading around subcarrier 10, but note particularly *all* the subcarriers in the vicinity of subcarrier 10 are faded as well. In general, the fading applied to nearby subcarriers will tend to be similar. Since modulated symbols are assigned to subcarriers in order, the result will be a burst of symbols with poor signal-to-noise ratio applied to the decoding algorithm, or a burst of errors.

Similarly, depending on the type of hopping applied to the current packet, we may see similar fade characteristics from hop to hop. In Figure 3.33, we see the situation where the center frequency hops between symbols. Since the new symbol is on an entirely new center frequency, we would expect, as shown in the figure, that there is little relationship between the fading on one symbol and that on the next. However, in the case where the center frequency

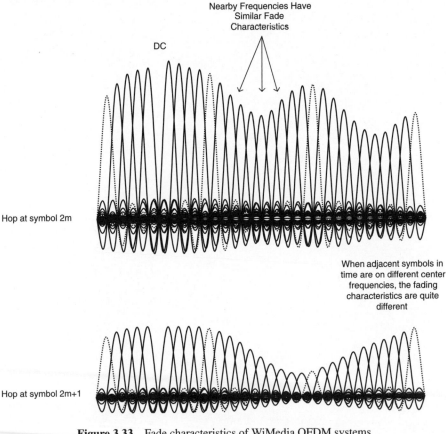

Figure 3.33   Fade characteristics of WiMedia OFDM systems

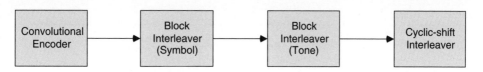

Figure 3.34   Interleaving in the WiMedia PHY

does not hop between symbols, the fading on a given subcarrier within one symbol and the same subcarrier of the next symbol will be nearly identical.

These two types of correlation between faded subcarriers determine the structure of the interleaving used to manage fade bursts in WiMedia PHY. There are three interleavers used in the WiMedia PHY: two standard block interleavers and a cyclic-shift interleaver. The coding and interleaving process is shown in Figure 3.34. The first interleaver, the symbol interleaver, is intended to manage the symbol-to-symbol correlation shown in Figure 3.35. In this case, the coded bits for an entire hop frame are placed in a rectangular buffer, as shown in Figure 3.35. In the modes where there is time spreading (modes 0 and 1, or data rates 0, 1, 2, 3, 4) there are three OFDM symbols per hop frame, and when there is no time spreading there are six OFDM symbols per hop frame. The depth of the interleaver is the number of OFDM symbols per hop frame. Since the buffer is rectangular, the width of the buffer must be the number of coded bits per OFDM symbol. Table 3.11 shows the width and depth of the interleavers for all of PHY layer modes.

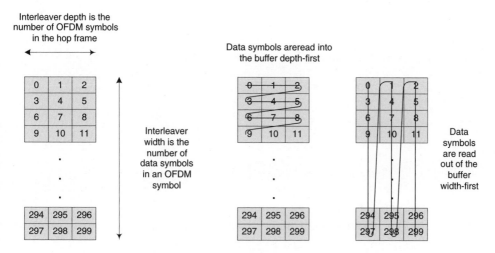

Figure 3.35   Symbol interleaver for data rates 0 and 1

Table 3.11 Interleaver parameters

Mode	Symbol interleaver		Tone interleaver		Cyclic shift
	Depth	Width	Depth	Width	
0	3	100	10	10	33
1	3	200	10	20	66
2	6	200	10	20	33

The bits are placed in the buffer depth-first in natural order. They are read out width first. This results in a coded-bit order that is 'mixed up' over symbols. Note that, for the last hop frame in a packet, it is possible that there will not be enough bits to fill an interleaver buffer. In this case, *pad bits* are added. The pad bits are added before coding and will be described later. The objective of pad bits is to ensure that the symbol interleaver is completely filled.

The tone interleaver is exactly the same in concept, but the width and depth are different. In this case, we interleave all the coded bits in a single OFDM symbol. The tone interleaver depth is always 10 and the width adjusted to ensure that a full symbol's worth of coded bits is placed in the interleaver. Since we have already added pad bits to fill out a hop frame, there is no further need for padding. The tone interleaver simply ensures that bits that are presented to the channel sequentially are not presented sequentially to the decoder. This makes sure that errors are not presented to the decoder in bursts.

The final step in the interleaving process is a cyclic shifter. The cyclic shifter also operates on a hop frame of data, one symbol at a time. Figure 3.36 shows the operation of the shifter for Mode 0. The first column of numbers is the relative positioning for the first symbol in the hop frame. The second column shows that the bits for the second symbol have been shifted by 33 bits, and the third column shows that the bits have been shifted by 66. The cyclic shifter simply ensures that the fading characteristics vary from symbol to symbol.

3.4.5 Summary

In this section we have discussed the error correction and related features: all of the bit-level processing performed by PHY transmitter. The input and the output of the processing described in this section are bits, whereas the input to the modulation section is bits and the output is a digitized waveform. Figure 3.37 shows the general process from beginning to end. There are still some details that we will discuss as we assemble a packet in the next section.

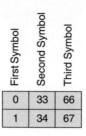

Figure 3.36   Cyclic shifter

3.5 Packet Structure

So far, we have discussed all of the machinery that is required to build and transmit a PHY packet: a PPDU. What remains is to walk through the details of the packet construction, carefully specifying the elements of the packet and the mechanisms used to create each element.

ECMA-368 specifies that the PHY connection be able to work in two modes: standard mode and burst mode. There are minor differences in both packet construction and timing between the two modes. Let us examine those differences.

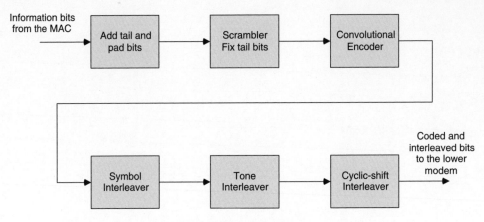

Figure 3.37   Outer modem

The structure of the PPDU is shown in Figure 3.4. Each packet will have a preamble, a header, and a payload. Figure 3.38 shows the top-layer view of the construction of a PPDU. The MAC provides the PHY with the necessary elements of the PPDU, including:

- The PSDU, or the actual data to be transmitted. This may actually be empty for some packet types (e.g. Immediate Acknowledgement packets), since the MAC sublayer header may contain all of the information that is to be conveyed to the receiving device. Along with the payload data, the MAC will provide a 32-bit FCS.
- A structure called TXVECTOR is simply the information the PHY needs to form the PLCP header (including PHY and MAC headers), and to transmit the entire packet. TXVECTOR contains the length of the packet, the payload data rate, and other critical information, and is shown in Table 3.12. Note that the MAC header is part of TXVECTOR.

3.5.1 Standard and Burst Modes

The PHY specification supports two transmission modes: Standard mode and Burst mode. In Standard mode, successive packets are separated with a Short Inter-Frame Spacing (SIFS) duration, i.e. 10 μs. This mode is mandatory to implement and gives enough time for any device to turn around from receive (RX) to transmit (TX) in between packets. Because (a) successive packets may come from different devices, (b) timing is not coordinated among the transmitting devices, and (c) the environmental condition may be different for packets received from different transmitters, the receiver must be prepared to *detect* the packet and must assume

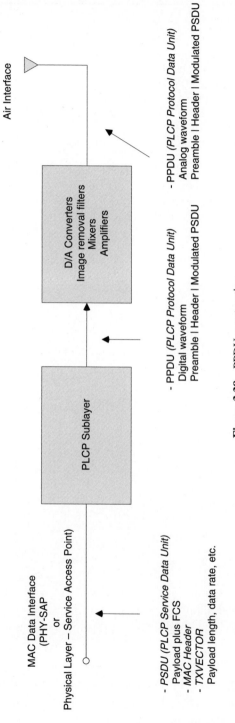

Figure 3.38 PPDU construction

MAC Data Interface
(PHY-SAP
or
Physical Layer – Service Access Point)

- PSDU (PLCP Service Data Unit)
  Payload plus FCS
- MAC Header
- TXVECTOR
  Payload length, data rate, etc.

PLCP Sublayer

- PPDU (PLCP Protocol Data Unit)
  Digital waveform
  Preamble I Header I Modulated PSDU

D/A Converters
Image removal filters
Mixers
Amplifiers

- PPDU (PLCP Protocol Data Unit)
  Analog waveform
  Preamble I Header I Modulated PSDU

Air Interface

Table 3.12 TXVECTOR elements

Parameter name	Description
Length	The number of octets in the payload. This length does *not* include the FCS, and ranges from zero to a maximum of 4095
Data rate	The rate at which the PSDU portion of the packet is to be transmitted
Burst mode	Informs the PHY layer that the *next* packet is sent in Burst mode, i.e. that the next packet will follow this packet with only one hop-frame (six symbols) of dead air between packets (also known as MIFS). See Section 3.5.1 for more detail
Preamble type	Indicates the preamble type (burst or standard) of the *next* packet. Valid only in Burst mode (see Section 3.5.1)
Scrambler seed	The final two bits of the scrambler seed, so that the receiver and transmitter scramblers may be synchronized
TX power	A number specifying the transmit power level
TFC and band group	Specifies the channel (TFC and band group) to transmit on. Note that only the least-significant bit (LSB) of the band group is specified, so PHY will need to obtain the full band group information from a different path
MAC header	10 bytes of MAC header information
Null tones	A bit field specifying nulled tones

that all timing and phase parameters need to be estimated. In other words, no a priori knowledge about the transmitter and the channel condition may be assumed from packet to packet. Thus, in Standard mode, standard PLCP preamble must be used, as opposed to burst PLCP preambles. Standard preambles are the full version of the preamble with a duration of 9.37 μs. (Burst preambles are about half as long, i.e. 5.625 μs.)

In Burst mode, on the other hand, a sequence of packets will be sent from a single transmitter. This means, the inter-packet timing is expected to be constant and, more importantly, no TX–RX turnaround time is required. Thus, in this mode, a shortened inter-frame spacing is used, called the Minimum Inter-Frame Spacing (MIFS), to reduce PHY overhead. The MIFS duration is exactly one hop-frame (six OFDM symbols) in duration, i.e. 1.875 μs.

Moreover, in Burst mode, the transmitter is allowed (for rates above 200 Mbps) to place a shortened PLCP preamble – burst preamble – instead of the standard preamble. The burst preamble further reduces PHY overhead and increases the throughput. The receiver should already have timing and phase parameters from the previous packet in the burst and, therefore, should be able to make the detection and channel estimation using the short preamble with sufficient accuracy. The PHY specification does not require phase coherence for either RF or sample clock timing. The MIFS packet spacing requirement has at least two advantages. First, it simplifies packet detection. The receiver knows exactly when to expect the next packet. Second, it is reasonable to assume that the channel will have changed only slightly between packets, so the estimation problem is simplified even if it is not eliminated.

Care must be taken not to confuse Standard and Burst modes of PHY with standard and burst preambles. They are not synonymous. Only standard preamble may be used in the Standard mode. Both standard and burst preambles may be used in the Burst mode. However, all transmissions at or below 200 Mbps must use standard preamble, whether in Burst mode or not. Also, the first packet of every burst of packets in the Burst mode must use standard preamble.

Burst mode is intended to be used for data transmission only, so the standard requires that each Burst mode packet have at least one byte of data. In Standard mode, the payload may be empty (for certain control packets, such as Immediate Acknowledgement), in which case no modulated PSDU will be transmitted.

3.5.2 PLCP Preamble

Since the WiMedia system is a multiuser, packet-based system, and individual nodes are neither time- nor frequency-coordinated, the receiver will need to perform at least the following for each packet:

1. *Packet detection*. Determine the presence of a packet.
2. *Timing estimation*. Determine when symbols and packets start.
3. *Phase estimation*. QPSK and DCM require phase-coherent reception.
4. *Amplitude estimation*. DCM requires amplitude estimates as well.

The first segment of each PPDU (packet) is a preamble and is designed to simplify the receiver's task for the items above. The preamble contains no information. It is a predetermined sequence of samples.

The PLCP preamble is composed of two segments: the synchronization sequence and the channel estimation sequence. The former is intended to simplify the detection of packets and to ensure that the receiver is able to synchronize timing, both symbol (hop-frame) timing and sample timing. The channel estimation sequence is intended to provide a simple mechanism for estimating the phase and amplitude of each subcarrier.

Unlike some other standardized OFDM systems, the synchronization sequence in ECMA-368 is not an OFDM waveform. It is created from a series of predefined waveforms, tabulated in detail in the standard. There are 10 predefined waveforms,

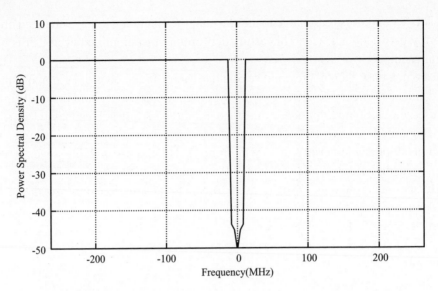

Figure 3.39   Spectral properties for the base time-domain sequence for TFC 1

called base time-domain sequences, one sequence for each TFC. The 10 sequences
were designed with three properties:

- Spectral properties similar to an OFDM waveform, including a DC null to allow for DC management in the receiver. Figure 3.39 shows the DFT for TFC 1.
  The flat spectrum allows the preamble to be transmitted at the maximum power,
  and the preamble should be transmitted at the same power level as the OFDM
  symbols.
- Good autocorrelation properties. Each of the sequences exhibits a single strong
  autocorrelation peak, with reasonably low side-peaks. Figure 3.40 shows the autocorrelation properties for TFC 1. The figure shows the output of a correlator, with strong peaks occurring when the input is lined up with the correlation
  sequence.
- Reasonably low cross-correlation between the different base sequences. Figure
  3.41 illustrates the cross-correlation waveform between the base sequences for
  TFC 1 and TFC 2. The correlation properties make a correlation detector practical and allow such a detector to distinguish devices transmitting on two separate
  channels.

When the ZP is added to the base time-domain sequence and the cover sequence
applied, the power spectrum is not nearly as flat as that depicted in Figure 3.39.
Furthermore, regulatory compliance requires measurement over a long period relative to the symbol time. With this in mind, the standards body has created a set

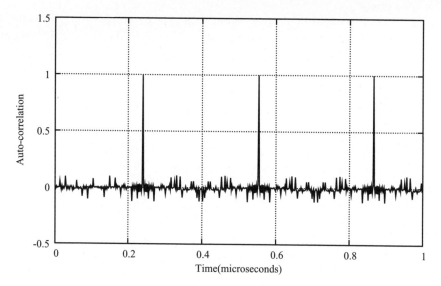

Figure 3.40   Autocorrelation properties for the base time-domain sequence for TFC 1

of seven alternate base time-domain sequences that correlate very well with the respective base time-domain sequences but have better spectral properties when measured according to most regulatory requirements. The correlation property allows a single correlator to be used for either the base sequence or the alternate sequence.

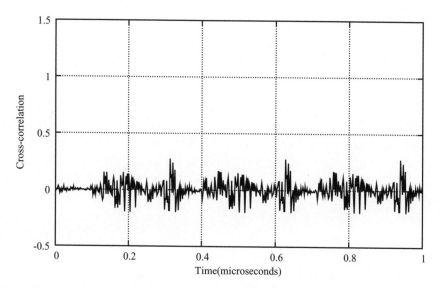

Figure 3.41   Cross-correlation between the base time-domain sequences with ZP for TFC 1 and TFC 2

Each of the 17 base time-domain sequences is 128 samples long, just like an OFDM symbol. A ZP of 37 zeros is added to the end of the base sequence. The base sequence plus ZP is duplicated in time, and then each duplicated element multiplied by the cover sequence value of ± 1. The standard preamble is 24 symbols, or duplications of the base sequence, and the burst preamble is 12 symbols long.

Figure 3.42 shows the process for constructing the synchronization preamble. The resulting sequence may then be correlated with the base sequence for the TFC the receiver is attempting to detect. The sign changes introduced by the cover sequence provide the receiver with *frame synchronization* reference points. For a receiver using typical algorithms, the synchronization sequence will be used for packet detection, coarse symbol timing, carrier frequency estimation, and frame synchronization.

The second segment of the preamble is the channel estimation sequence. The channel estimation sequence is six identical OFDM symbols plus ZP using a set of predefined inputs. Since the receiver knows the modulation values, it is able to determine amplitude and phase for each subcarrier, at least to within some estimation error. The channel estimation sequence is typically used for amplitude and phase estimation, plus fine frequency and symbol timing. Figure 3.43 shows the construction process for the channel estimation sequence.

Finally, the PLCP preamble is created from the two subsequences. The TFCs are applied, beginning with the first symbol in the preamble, as shown in Figure 3.44. Both the synchronization preamble and the channel estimation sequence are real sequences; the imaginary part is zero. The preamble must be placed in the real part of the transmitted waveform.

3.5.3 PLCP Header

As shown in Figure 3.45, the PLCP header contains two short segments of information: the PHY header constructed by the PHY, and the MAC header provided to the PHY by the MAC sublayer. The PLCP header is always transmitted using Mode 1 modulation parameters from Table 3.8 and uses the rate 1/3 convolutional code. That is, the PLCP header is transmitted at the lowest PHY data rate for maximum reception reliability. The PLCP header is always two hop-frames long, or 12 symbols. In this section we walk through the construction of the PLCP header.

The PHY constructs a PHY header from the information provided by the MAC sublayer in the TXVECTOR (see Table 3.12). The PHY header contains:

- the rate index (see Rate ID in Table 3.8) for the data rate at which the payload section will be transmitted;
- the number of octets in the payload;

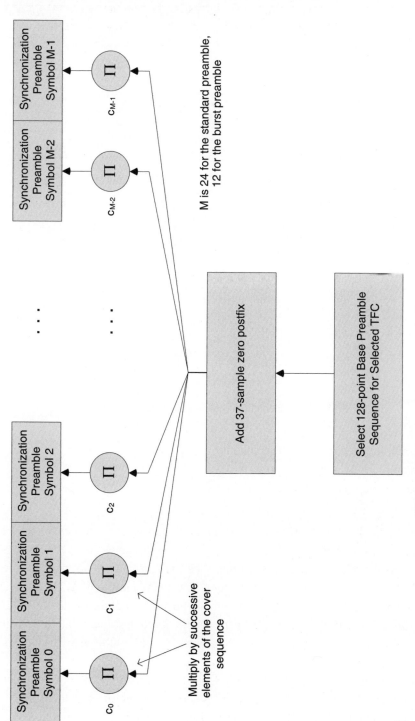

Figure 3.42 Construction of the synchronization preamble

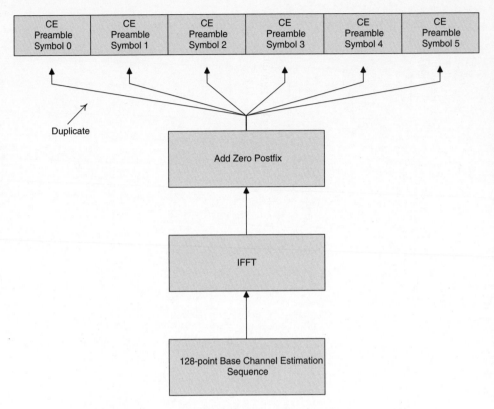

Figure 3.43   Construction of the channel estimation (CE) preamble

- the scrambler seed used by the transmitter;
- the preamble type and burst mode;
- The band group (LSB only) and TFC used to transmit the packet.

The actual encoding, bit by bit, is in the standard and not included here. The PHY header is 40 bits, or 5 bytes, long.

The MAC header is 10 bytes, or 80 bits. The PHY layer does not need to know the structure or content of the MAC header. The PHY and the MAC headers are included in the PLCP header since (a) neither header is part of the payload, (b) both headers are critical to the decoding of the payload and are thus modulated using the most robust techniques used in the PLCP header, and (c) the MAC header may be the only information conveyed by the packet (zero payload).

The PLCP header uses elements from Section 3.4. First, a stream of bits is created from the PHY and the MAC headers. Selected scrambling and then the HCS are applied. Next, the Reed–Solomon parity bits are added, and finally the result is encoded with the convolutional code.

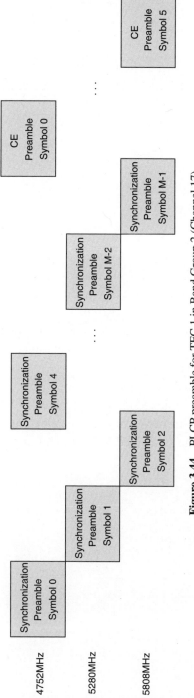

Figure 3.44 PLCP preamble for TFC 1 in Band Group 2 (Channel 17)

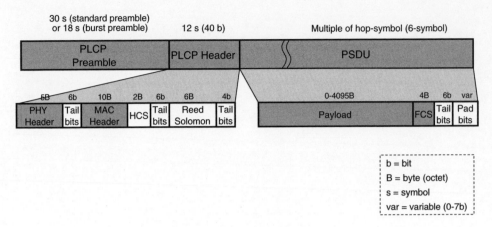

Figure 3.45   PPDU format

Figure 3.46 shows the details of the construction of the PLCP header and Figure 3.47 the specifics of creating the bit stream before it is convolutionally encoded. The reader should note that a detailed example of header construction is provided in Annex E of ECMA-368. First, the PHY header is created from the TXVECTOR information. The MAC and PHY headers are used together to construct the HCS. Protecting both headers with the HCS is critical, since loss of either will most likely cause the rest of the packet to be lost.

The MAC header and HCS are scrambled. The PHY header contains the scrambling seed, so it is not scrambled since the receiver will not be able to interpret the scrambling seed field from a scrambled header. The resulting unscrambled PHY header along with the scrambled MAC header and HCS are encoded using

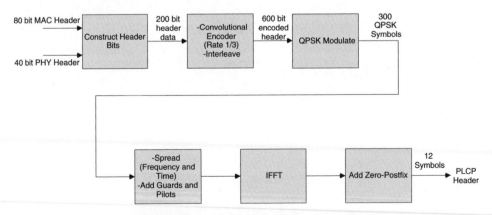

Figure 3.46   Construction of PLCP Header

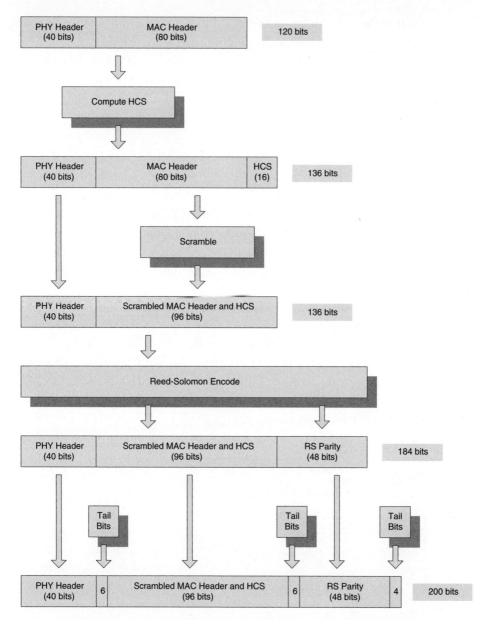

Figure 3.47   Construction of the header bits before convolutional encoding

the Reed–Solomon encoder. The outcome is the uncoded bits along with a 6-byte parity field appended to the headers. Six tail bits are added at the end of the PHY header, as well as the combination of the scrambled MAC header and HCS. Four tail bits are added at the end of the packet. This allows the decoder to operate on each of the subfields independently, slightly increasing the reliability for the entire packet. Of course the four tail bits at the end of the header are not enough to drive the encoder to the zero state, but they certainly help.

Finally, the resulting 200 bits are encoded with the rate 1/3 convolutional encoder and then interleaved. The 600 coded and interleaved bits are QPSK modulated, spread, guard and pilots added all according to Section 3.2, and the final PLCP header is computed using the IFFT.

3.5.4 Modulated PSDU

The final segment of the PHY packet or PPDU is the modulated PSDU. In the WiMedia PHY specification, the term PSDU is somewhat ambiguously defined and is used for both the octet stream passed to the PHY layer by the MAC sublayer and for the OFDM modulated waveform transmitted by the device.[2] The information content is the same in both cases; but, since the first form of the PSDU is a sequence of bits and the second is a waveform, we will be a bit more specific. First, the PSDU in our terminology will be the payload provided by the MAC sublayer. The waveform as presented by the PLCP sublayer to the PMD sublayer is, in our architectural view, a digitized waveform consisting of a sequence of symbols, or an OFDM modulated waveform. We will term this waveform the *modulated PSDU*. As the modulated PSDU passes through the elements of the PMD, it will become an analog waveform. We will still call it the modulated PSDU, but we recognize that the analog elements will add distortion.

Figure 3.48 shows the details of the construction of the PSDU and Figure 3.49 the specifics of creating the bit stream before it is convolutionally encoded. The payload and FCS as provided by the MAC are concatenated by the PHY, and six zero tail bits are appended. The interleaving requires a complete hop-frame, so pad bits are added before any further processing. The number of pad bits is simply enough zero bits to ensure that the scrambler has an integer number of information bits per hop frame ($\bar{N}_b$ in Table 3.8). The entire result is scrambled, but the scrambled tail bits are then replaced with zeros to ensure that the encoder will return to the zero state.

The scrambled output is then encoded using the convolutional encoder. The puncturing pattern is derived from the rate provided by the MAC. This creates a sequence of bits which is an integer multiple of $\bar{N}_c$ values (see Table 3.8). Interleaving is applied; then the outcome is modulated using QPSK or DCM depending on the data rate. Guards and pilots are inserted into the stream of data symbols, spreading is applied, and the result is OFDM modulated. Finally, ZP is inserted for each symbol.

---

[2] Note that there is no ambiguity in mapping the payload to the packet, just a minor terminology issue.

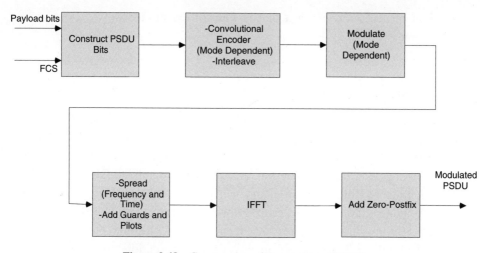

Figure 3.48   Construction of the modulated PSDU

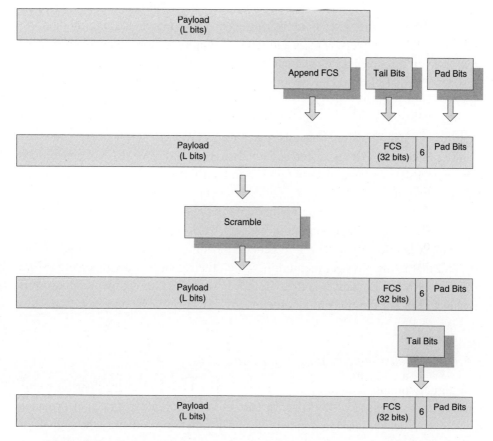

Figure 3.49   Construction of the modulated PSDU bits before convolutional encoding

The final result is a waveform that is transmitted immediately following the PLCP header.

3.6 Performance Requirements

We have specified the PPDU in some detail, but have only discussed it in idealized, mathematically precise terms. Furthermore, we have only specified the PPDU at the input to the PMD. As we have said before, the WiMedia standard is largely concerned with the air interface, or the transmitted form of the PPDU. To complete the specification, it will be necessary to specify not just the mathematical form of the PPDU, but also the amount of error allowed in the transmitted PPDU.

Figure 3.50 shows the complete transceiver system with some (but by no means all) of the possible error sources. Typically, the digital section introduces numerical errors, such as finite precision (quantization) and overflow. The analog section introduces many more. Data converters introduce (at a minimum) additional quantization, clipping, and other nonlinear effects. Amplifiers create intermodulation distortion and DC offsets. Mixers add intermodulation distortion and carrier leakage. Every transceiver design has unique characteristics, so it is not possible or reasonable to list all of the error sources and create a specification of each. Like most standards, the WiMedia specification specifies the maximum aggregate error allowed. In this chapter, we will discuss the WiMedia error clauses and relate them to some simple signal-to-noise ratio terms. We will further define the link quality indicator (LQI) and received signal strength indicator (RSSI), since those measures are also related to signal power and noise power. Finally, we end with a brief discussion of wireless medium and channel modeling.

In this section, we will be using the word *channel* not as it was used in previous sections, i.e. as a logical conduit for peer PHYs to talk through. Instead, here, the channel is referring to the more generic meaning used in communication theory, and that is the physical medium in which the signal will be traversing. More specifically, a UWB channel is the wireless environment through which the UWB signals will pass from transmitter antenna to the receiver antenna.

3.6.1 S/N Measurements and Terminology

In a digital communication system, the average ratio of the energy per bit to the noise spectral density E_b/N_0 is the primary reference for bit error rate performance measures. When discussing acquisition and estimation problems, the signal-to-noise ratio S/N is more convenient. In OFDM systems, we may relate S/N values, error vector magnitudes (EVMs), and other measures to E_b/N_0. This subsection

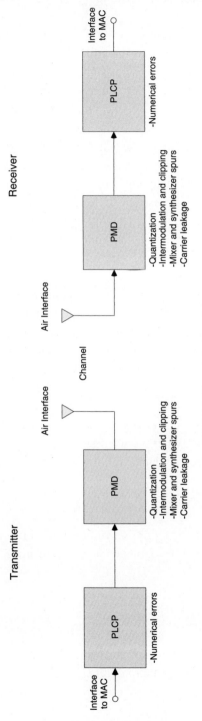

Figure 3.50 PPDU error sources

is a brief description of some basic terminology that we will use throughout this section.

For our mathematical descriptions, we will assume that the noise is white and that the signal level is constant across the OFDM spectrum. The results are easily generalized to nonwhite and nonconstant signals, and, more importantly, it must be understood that the WiMedia PHY requirements are applied to the more complex case. However, this is simply a conceptual introduction, and the additional complexity is not warranted.

There are many factors that make the constant E_b/N_0 assumption questionable, including filter ripple and channel response. In fact, in a UWB system, N_0 is not likely to be constant since RF matching across the entire band is not perfect. Other error sources contribute to the noise spectral density, including quantization error and other distortions. Some, such as quantization error, are fairly white, and others, such as clipping and intermodulation distortion, contribute nonwhite components. In spite of all this, the assumptions of constant signal and white noise are reasonable for a first-order approximation as long as we assume a flat channel response, which is almost always the case when making compliance measurements.

There are a number of signal-to-noise quantities that may be of interest. Of course, the energy per bit is the most useful concept when examining error rate performance, but other measures are more useful when examining the performance of data converters or estimation algorithms. We will define several useful quantities and relate them to each other. Figure 3.51 shows the energy quantities in their respective locations in the receive process.

We will introduce the *energy per tone* E_t as the energy in a single tone or subcarrier. To compute the energy per bit, we first despread by the spreading factor S, yielding the energy per received data symbol:

$$E_s = SE_t$$

The demodulation maps a single QPSK symbol into 2 bits or two DCM symbols into 4 bits, so that there are $N_b = 2$ bits per symbol, resulting in energy per coded bit of

$$E_c = \frac{E_s}{N_b} = \frac{SE_t}{N_b}$$

Finally, the decoding process yields R_c bits (or information bits) per code bit, so that the energy per received information bit is

$$E_b = \frac{E_s}{R_c} = \frac{SE_t}{R_c N_b}$$

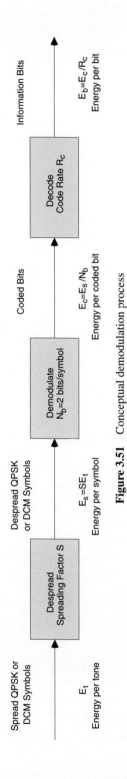

Figure 3.51 Conceptual demodulation process

This allows us to express E_b/N_0 in terms of E_t/N_0:

$$\frac{E_b}{N_0} = \frac{S}{R_c N_b} \frac{E_t}{N_0}$$

We may also compute OFDM signal quantities from the value of E_t/N_0. The energy in a single OFDM symbol is

$$E_o = N_t E_t$$

where $N_t = 122$ is the number of active tones in the WiMedia waveform so that the average power in the OFDM waveform is

$$P_o = \frac{E_o}{T_s} = \frac{N_t E_t}{T_s}$$

Note that we are assuming a flat channel with white Gaussian noise for all of these quantities. Further, there is no difference in energy between an OFDM symbol and a symbol, since the ZP contributes no energy. Now, if B_n is the equivalent noise bandwidth of the system, then the noise power is

$$P_n = B_n N_0$$

allowing us to write the S/N as

$$S/N = \frac{P_o}{P_n} = \frac{N_t}{T_s B_n} \frac{E_t}{N_0}$$

3.6.2 Transmitter Error Specifications

We have already discussed some of the transmitter specifications. In particular, the transmitter center frequency and the symbol clock must be accurate to within 20 ppm and must be derived from the same source. In this section we will discuss power spectrum limitations and constellation errors.

3.6.2.1 Transmit Spectral Mask

WiMedia PHY transmitters are required to keep emissions within the limits of a spectral mask with breakpoints shown in Table 3.13 and Figure 3.52. The mask is intended to ensure that signals in adjacent bands present minimal mutual interference. When comparing a transmitted signal against the mask, the unit dBr is

Table 3.13   WiMedia spectrum mask breakpoints

Frequency range (MHz) relative to transmitter center frequency f_c	Value (dBr) relative to peak spectral density
$f_c - 330$	−20
$f_c - 285$	−12
$f_c - 260$	0
$f_c + 260$	0
$f_c + 285$	−12
$f_c + 330$	−20

usually used. dBr refers to the power (in decibels) relative to the peak power in the mask. Thus, 0 dBr is the peak measured value of the spectral density. The measured power spectrum must remain below the mask at all points. The ECMA-368 document does not specify resolution bandwidth or other measurement parameters. The reader is cautioned that some regulatory bodies have stricter requirements than the PHY specification does.

It can be shown that the power spectrum of a perfect OFDM signal, without noise or other distortions, and with each subcarrier having the same energy per

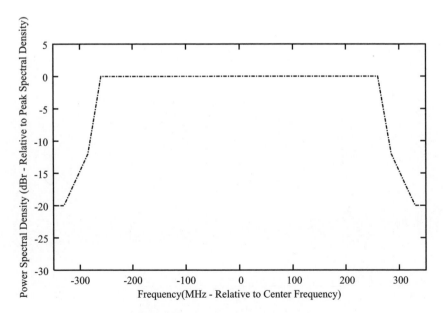

Figure 3.52   Transmit spectrum mask

tone E_t, may be expressed as

$$S_o(f) = \frac{E_t}{2} \frac{T_u}{T_s} \sum_{\substack{k=-61 \\ k \neq 0}}^{61} \frac{\sin^2[\pi(f - f_k)T_u]}{[\pi(f - f_k)T_u]^2}$$

provided that the FFT inputs $X_m[k]$ are uncorrelated for all m and k [6]. In the WiMedia system, $T_u = 128\Delta t$ and $T_s = 165\Delta t$, where Δt is the sample interval defined in Table 3.4. The peak spectral density is

$$\max S_o(f) = \frac{E_t}{2} \frac{T_u}{T_s}$$

so that the relative spectral density is

$$S_o(f) = \sum_{\substack{k=-61 \\ k \neq 0}}^{61} \frac{\sin^2[\pi(f - f_k)T_u]}{[\pi(f - f_k)T_u]^2}$$

Figure 3.53 shows the ideal spectral density compared with the spectral mask for $E_t = 1$ and Figure 3.54 shows a system with some nonlinearity in the analog section. The ideal spectral density easily meets the mask, but the spectral density after the nonlinear distortion does not.

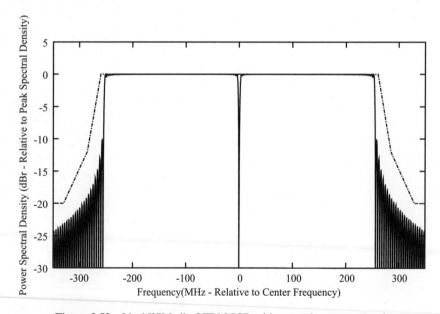

Figure 3.53   Ideal WiMedia OFDM PSD with transmit spectrum mask

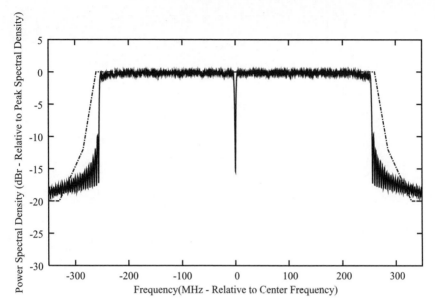

Figure 3.54   WiMedia OFDM PSD with nonlinear distortion and transmit spectrum mask

3.6.2.2 Transmitter Constellation Error

In addition to the transmit spectrum, and other transmit requirements discussed already, the WiMedia PHY specification requires that the transmitter constellation be reasonably accurate in terms of distance from a center point. Figure 3.55 shows a typical QPSK transmitter constellation when the only error source is quantization error. The horizontal axis is the real part of the FFT output and the vertical axis is the imaginary part, and the 'balls' are the FFT outputs for a large number of symbols. Other types of distortion, such as a poor synthesizer, may case the constellation to smear in various directions. Nonlinear distortions may result in nonwhite noise components as well as compression of the constellation. Other transmit imperfections distort the constellation in different ways.

The WiMedia PHY specification requires that transmitter constellation errors be measured using a form of EVM. The EVM measures the Euclidean distance between the complex numbers observed by a receiver and the actual center of the constellation. The EVM is intended to be measured with a near-ideal receiver. Since the measurement is made at the transmitter, there will be little thermal noise, so that any good receiver should work for EVM measurements.

The ECMA-368 definition for EVM appears somewhat mysterious at first, and assumes the use of the overlap-add operation in the receiver in spite of the fact that the overlap-add technique is never defined in the standard. The WiMedia EVM

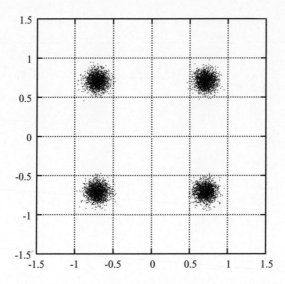

Figure 3.55   Constellation in quantization noise and with no filtering

definition is

$$\varepsilon = \frac{1}{N_f} \sum_{i=1}^{N_f} \sqrt{\frac{1}{N_{frame}} \sum_{n=N_{sync}+N_{hdr}}^{N_{packet}} \left[\frac{1}{P_o} \frac{1}{N_D + N_P} \sum_{k=1}^{N_D} \left| R_{D,n}[k] - C_{D,n}[k] \right|^2 + \sum_{k=1}^{N_D} \left| R_{P,n}[k] - C_{P,n}[k] \right|^2 \right]}$$

We will dissect the EVM equation term by term. The outer sum is simply the average of the term in the radical over N_f packets, where a minimum of 100 packets is required. The sum under the radical is the average over the payload (modulated PSDU) portion of the packet or PPDU. A minimum of 30 symbols must be included in the PSDU.

The terms inside the inner summation

$$\left| R_{D,n}[k] - C_{D,n}[k] \right|^2$$

and

$$\left| R_{P,n}[k] - C_{P,n}[k] \right|^2$$

are the normed differences between the 'observed subcarriers' and the 'transmitted constellations' for the pilot and data symbols (but not the guards). The observed subcarriers are the values produced at the FFT output. We may model these terms as

$$R_m[k] = \sqrt{E_t[k]} X_m[k] + N_m[k]$$

where we have lumped the pilot and data symbols together using the terminology from Section 3.2. We have assumed that the energy per tone does not vary from symbol to symbol during a packet. The term $N_m[k]$ is the error vector, and is the term that the EVM technique is generally intended to measure. The error vector is not necessarily a normal (Gaussian-distributed) random variable, nor is it necessarily of zero mean. The transmitted constellation values, then, are simply the values $C_m[k] = \sqrt{E_t[k]}X[k]$, or the IFFT inputs scaled by the amplifiers, filters, etc., in the transmit chain. Thus, we see that

$$\frac{1}{N_D + N_P} \sum_{k=1}^{N_D} \left| R_{D,n}[k] - C_{D,n}[k] \right|^2 + \sum_{k=1}^{N_D} \left| R_{P,n}[k] - C_{P,n}[k] \right|^2$$

$$= \frac{1}{112} \sum_{\substack{k=-56 \\ k \neq 0}}^{56} |N_m[k]|^2$$

is the average value of the squared error vector. Averaging is done over data and pilot tones, over a single symbol.

The average power of the data and pilot constellations is

$$P_o = \frac{1}{112} \sum_{\substack{k=-56 \\ k \neq 0}}^{56} |C_n[k]|^2 = \frac{1}{112} \sum_{\substack{k=-56 \\ k \neq 0}}^{56} |E_t[k]|^2$$

We get an interesting interpretation of the EVM if we assume that the energy per tone is constant, that is if $E_t[k] = E_t$. If the error vector is zero mean with variance σ^2 as would be the case in AWGN, then we have

$$\frac{|N_m[k]|^2}{P_o} = \frac{1}{E_t/\sigma^2}$$

If the noise at passband is white and Gaussian with power spectral density N_0, then we have

$$\sigma^2 = \frac{N_0}{L_o}$$

where $L_o = N/(N + N_{ZP})$ is the overlap-add processing loss incurred by adding noise samples to the active portion of the symbol. Thus, in AWGN and a nondispersive channel, we have

$$\varepsilon = \frac{L_o}{E_t/N_0}$$

The WiMedia PHY specification sets limits on the EVM to ensure that transmitter degradation will not have a significant impact on receiver performance. Since

Table 3.14 Permissible EVM values

	EVM (dB)		
Data rate (Rate ID)	No TX attenuation	TX attenuation of 2, 4, 6 dB	TX attenuation of 8, 10, 12 dB
0, 1, 2, 3, 4	−17.0	−15.5	−14.5
5, 6, 7	−19.5	−18.0	−17.0

lower data rates are less susceptible to noise, and use only QPSK, their EVM limits
are lower than those of the higher data rates that use DCM. The standard further
allows for EVM degradation with power control, since lower transmit power is
normally used in better channels. Table 3.14 shows the worst-case EVM values
permitted for a compliant transmitter. Note that the WiMedia standard uses the
term 'relative constellation error' instead of EVM.

3.6.3 Receiver Performance Specifications

The WiMedia PHY specification requires that the receiver sensitivity meet (be
below) the values shown in Table 3.15. The receiver sensitivity, in this standard,
is defined as the minimum power required at the input of the receiver to achieve a
packet error rate of 8 % for packets of 1024 octets in Band Group 1. (In other band
groups, a 1–2 dB noise figure degradation is allowed.)

The receiver sensitivity is dependent on the data rate, receiver noise figure, and
any implementation losses. It is interesting to examine the receiver performance
under the conditions required to meet sensitivity. First, Table 3.16 details the parameters required to make signal and noise computations. We have already shown
that the signal power and energy per tone are related by

$$P_o = \frac{N_t E_t}{T_s}$$

Table 3.15 Receiver sensitivity

Data rate (Mbits/s)	Minimum receiver sensitivity (dBm)
53.3	−80.8
80	−78.9
106.7	−77.8
160	−75.9
200	−74.9
320	−72.8
400	−71.5
480	−70.4

Table 3.16 Data-rate parameters for signal and noise power computations

Data rate (Mbits/s)	Spreading factor S	Code rate R_c	Coding gain G_c (dB)
53.3	4	1/3	5.7
80	4	1/2	5.2
106.7	2	1/3	5.7
160	2	1/2	5.2
200	2	5/8	4.9
320	1	1/2	5.2
400	1	5/8	4.9
480	1	3/4	4.5

and the energy per bit is related to the energy per tone via

$$E_b = \frac{S E_t}{R_c N_b}$$

The noise spectral density is approximately

$$N_0 = k T_{sys} F$$

where k is Boltzmann's constant, $T_{sys} = 290$ K is the noise temperature of the receiver, and $F = 6.6$ dB is the receiver noise figure. From this, and assuming that the noise bandwidth of the receiver is $B_n = 528$ MHz, we may compute the values in the first columns of Table 3.17. The column 'E_t/N_0 required for $P_b = 10^{-5}$' is computed by recognizing that a packet error rate of 8 % corresponds to $P_b = 10^{-5}$ for packets of 1024 octets. Both QPSK and DCM require an E_b/N_0 value of 9.59 dB for $P_b = 10^{-5}$ in an uncoded system. The coded system requires

$$\frac{E_b}{N_0} = \frac{E_b}{N_0}\bigg|_{uncoded} - G_c + 10 \log_{10}\left(\frac{N}{N + N_{ZP}}\right)$$

Table 3.17 Receiver performance at sensitivity

Data rate (Mbits/s)	Input signal power (dBm)	E_t/N_0 (dB)	E_b/N_0 (dB)	S/N (dB)	E_t/N_0 required for $P_b = 10^{-5}$ (dB)
53.3	−80.8	0.7	8.4	−0.7	−2.9
80	−78.9	2.6	8.6	1.3	−0.7
106.7	−77.8	3.7	8.4	2.4	0.1
160	−75.9	5.6	8.6	4.3	2.4
200	−74.9	7.0	9.0	5.7	3.6
320	−72.8	8.7	8.7	7.4	5.4
400	−71.5	10.0	9.0	8.7	6.6
480	−70.4	11.1	9.3	9.8	7.8

where the last term is the overlap-add loss described previously. Table 3.17 shows that the *S/N* is very low for the lowest data rates, presenting interesting problems for the various estimation algorithms required for accurate packet reception.

3.6.4 Signal Quality Measurements

When a packet is received by the PHY, there are two measurements of signal quality that are returned to the MAC in the RXVECTOR: the RSSI and the LQI. The PHY is required to implement the LQI, but is not required to implement the RSSI.

The RSSI is a number between 0 and 255, and is simply an indication of the signal energy at the received antenna. The WiMedia specification does not specify a mapping between the actual energy and the RSSI, but merely states that the RSSI is a monotonic function of the energy.

The LQI is more precisely defined than the RSSI. The receiver will make an estimate of *S/N* after the FFT (called the link quality estimate (LQE)). The standard does not specify how this measurement is to be made, nor does it precisely define *S/N* in this context, but simply that the LQE be 'nominally equal to the *S/N* after the FFT.' For devices that do not implement DCM, the LQE must report *S/N* values between −2 dB and +7 dB (see Table 3.17 for *S/N* values at sensitivity for these rates). Similarly, if DCM is implemented, then the LQE must report *S/N* values between −2 dB and +12 dB. The manufacturer may implement a larger *S/N* range, but it is required to test the LQE in a fixed AWGN channel with the accuracy shown in the standard, and the LQE is expected to be monotonic with input power.

LQI is an 8-bit number with 0 indicating that no valid measurement is possible, 1 indicates −6 dB, 2 indicates −5 dB, etc. The standard provides for vendor-specific LQI reporting as well.

The RSSI and LQI may be used together by the MAC to make decisions regarding PHY parameter selection (e.g. data rate). Normally, the RSSI and LQI should be coupled – increasing RSSI should yield increasing LQI. However, a large RSSI and poor LQI is an indication of interference or other problems in the channel.

3.6.5 Channels and Channel Modeling

It is beyond the scope of this chapter to provide a detailed exposition of channel modeling, but a brief description of the UWB channel and the techniques used to model it are worth a short discussion.

We have already shown that the extremely wide bandwidths associated with UWB create a serious problem with frequency selective fading. Molisch *et al.* [7] presented a model of the UWB channel, and this model has been used to evaluate the performance of UWB systems. There are four instantiations of the model that are consistently used by the WiMedia Alliance members in UWB system evaluation, termed CM1 (channel model 1), CM2, CM3, and CM4. CM1 is generally a mild multipath channel and CM4 a severe one. These instantiations are based on a single model with different parameters for random number selection. We will not describe these in detail, referring the reader to [7] instead. However, there are some useful observations to be made:

- In all of these models, if the 'shadowing' parameter is turned off, then the distribution of the magnitude of a particular FFT output is approximately a Rayleigh random variable.
- The correlation between the channel response of nearby FFT outputs is fairly high in CM1 and it is very low in CM4. This has many implications, one of which is that, using the WiMedia OFDM waveform in CM1, each subcarrier will have very little frequency-selective fading, but there may be some frequency-selective fading on each subcarrier when using CM4. This correlation property has an impact on some of the receiver estimation algorithms as well.
- In CM1 there is little ISI when the ZP is included, but in CM4 there may be considerable ISI.

The plots describing the channel response in this chapter were made using CM1.

3.7 Ranging

Computing the distance of a target wireless device from a reference wireless device is called ranging. Since UWB signals have such a wide bandwidth, their time-domain resolution is quite high, making them suitable for very accurate range measurements – on the order of centimeters. In WiMedia UWB technology, the method of time of arrival (ToA) is used to enable range measurements. Simply put, a device measures (estimates) the time it takes to receive a signal from another device (Δt) and multiplies it by the speed of RF propagation (c, the speed of light) to get the range. The accuracy of such range measurements depends on:

- available signal bandwidth;
- clock accuracy;
- estimation accuracy of time of arrival.

Combining range measurements from multiple reference devices can lead to positioning, which is outside of the scope of this section. Here, we will give a brief overview of the two-way ranging (TWR) measurement mechanism provided in the WiMedia PHY specification. Note that this is an optional feature.

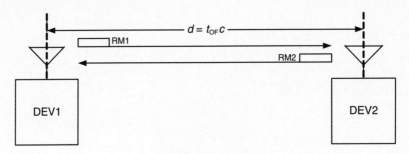

Figure 3.56 TWR

TWR based on ToA provides a viable ranging solution that does not necessitate clock synchronization among the devices in an ad hoc network. Let us consider a TWR that estimates the distance d between two devices as shown in Figure 3.56 by using the estimate of time of flight t_{OF} of the packet RM1 sent from Device 1 to Device 2 and the response packet RM2 sent back from Device 2 to Device 1. Thus, if c is the speed of propagation, then $d = t_{OF}c$.

Two-way ToA measurements do not require synchronized devices, since the round-trip propagation delay can be calculated at a single device and then divided by two to yield the estimate of the one-way time of flight. In general, ranging accuracy of the ToA-based method depends on the signal bandwidth used in the transactions. This is due to the fact that a signal with a higher bandwidth results in more accurate detection [8]. However, assuming perfect detection, the rate of the sampling clock affects the timing accuracy of the ranging transactions, i.e. the higher the rate of the sampling clock, the higher the ranging accuracy is. For instance, using a 528 MHz sampling clock rate gives the finite ranging resolution of 56.8 cm.

The transaction timing of the range measurement packets RM1 and RM2 related to TWR is illustrated in Figure 3.57. The absolute time when the RM1 leaves DEV1 is t_1 and the recorded counter value in DEV1 at time t_1 is T_1. That is, t_1 is

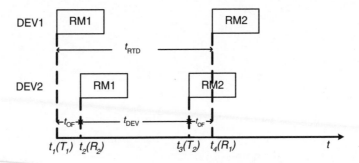

Figure 3.57 TWR transaction timing

the real time and T_1 is the PHY timing counter value representing t_1. At time t_2, RM1 reaches DEV2 and the recorded counter value in DEV2 at time t_2 is R_2. The processing time in DEV2 is t_{DEV} and the packet RM2 leaves DEV2 at time t_3 when the counter value in DEV2 is T_2. Thus, $t_{DEV} = t_3 - t_2$. RM2 reaches DEV1 at t_4 and the recorded counter value at time t_4 is R_1. Therefore, the round-trip delay is $t_{RTD} = t_4 - t_1$ and the time of flight t_{OF} is given by

$$t_{OF} = \frac{t_{RTD} - t_{DEV}}{2} = \frac{(t_4 - t_1) - (t_3 - t_2)}{2}$$

If all the timing instants are precisely recorded by the clock-counter value, then the time of flight can be expressed in terms of counter value as

$$t_{OF} = \frac{(R_1 - T_1) - (T_2 - R_2)}{2f_0}$$

where f_0 is the sampling frequency of the device clock of DEV1 and DEV2. Note that this equation represents the estimate of time of flight in terms of counter values or the measurements made in the devices and, therefore, realizable and does not need the time synchronization (absolute timing of different measurement instant) of the device clocks of DEV1 and DEV2. Since the distance between the two devices is given by $d = t_{OF}c$ and c is very large (3×10^{10} cm s^{-1} in free space), small errors in the estimate of time of flight will create large errors in distance calculation. There are various sources of such small errors. For instance, the sampling instant of the device clock is discrete; therefore, there will be an error between the actual time of the event and the time when a measurement is taken, or, recorded as counter value.

Theoretical but inaccurate assumptions that can be made in developing a TWR algorithm are:

- the clock frequencies in the two devices are exactly the same;
- there is no frequency drift in the clocks;
- there is no time gap between the generation (detection) and the transmission (arrival) of the reference signal;
- the detection of the arrived signal is perfect and detected in the next immediate tick of the sampling clock from its arrival instant.

In real life, none of these is valid and needs to be addressed in order to get the ranging accuracy desired. One of the ways that accuracy may be increased is given in the standard: increasing the clock frequency available for ranging in the PHY. In

Table 3.18 Theoretical ranging accuracy for different clock frequencies

Timer clock rate (MHz)	Theoretical ranging accuracy (cm)
528	56.8
1056	28.4
2112	14.2
4224	7.1

fact, the standard calls for the availability[3] of a ranging counter of 16–32 bits long that can clock at a rate of at least 528 MHz. This is the minimum timer clock rate. Higher, optional, rates are given at 1.056 GHz, 2.112 GHz, and 4.224 GHz. The theoretical ranging accuracy obtainable with each of these clock rates is given in Table 3.18. Although increasing the clock frequency can reduce the quantization errors in time measurements, it can considerably increase the complexity of the device implementation. Hence, it would be desirable to reduce quantization errors without resorting to higher clock rates. The paper by Khawza and Heidari-Bateni [9] gives one such method.

As should be obvious by now, the TWR mechanism requires the MAC sub-layer's involvement, since ranging packets with ranging information are being passed back and forth between the two devices. Section 4.13 describes the MAC-level view of TWR.

3.8 PHY Services and Interfaces

We have described the form of the PPDU and the mechanisms by which the PPDU is constructed. We have also discussed TFCs and channels. An important question to which we have only alluded so far is how the PHY is supposed to choose a channel, a data rate, or any of the other elements that make up a PPDU. The answer is very simple: the PHY layer does what the MAC sublayer tells it to do. It makes no decisions by itself. What the PHY does is to provide a series of *services* to the upper layers and a set of interfaces to provide these services.

The basic services that the PHY layer provides are as follows:

- *Transmit a packet*. The MAC provides the PHY with the information required to transmit, including data, channel, and modulation parameters, and the PHY performs the transmission.

---

[3] If the ranging option is implemented.

- *Receive a packet.* MAC instructs PHY to go into receive mode along with the channel to listen to; and the PHY keeps searching for the arrival of a PPDU and reports it to the MAC sublayer.
- *Clear channel assessment (CCA).* The MAC instructs the PHY to determine whether the specified channel is clear or currently has a transmission in progress.
- *Support for ranging (optional).* The MAC instructs the PHY to assist in making range measurements with another device. The PHY sends and receives the required packets, keeping accurate timing for them and reporting the timing to the MAC.

In this section, we will describe each of these services in some detail. Note that the standard does not require any specific form of the services and interactions, so the information presented in the following paragraphs, as well as the associated section in the standard, is simply informative. Our purpose in presenting the information here is simply to clarify the information in the standard.

Figure 3.2 showed a simplified architectural view of the WiMedia PHY along with the exposed interfaces. Recall that any such architectural views are for reference and exposition only; the standard does not require any specific structure or implementation. The PHY needs only to meet the air interface requirements to be compliant. The reference architectural information is valuable because it provides a context in which to describe the PHY operation. We will use the reference architecture, including block diagrams and interface descriptions, to describe the PHY services in detail.

In the introductory text to this chapter we showed that the PHY has two exposed upper interfaces. The upper interfaces are the places that the upper network layers have input to the PHY, and these layers are able to obtain information from the PHY. One of the interfaces is PHY SAP, the data interface. The services offered are given in Table 3.19. The other interface is the PLME SAP, normally used to set parameters into the PHY or to obtain information from the PHY. The services offered are given in Table 3.20.

Table 3.19 PHY SAP services

Service	Description
PHY-DATA	Transfer data between the MAC and the PHY
PHY-TX-START	Begin transmitting a packet
PHY-TX-END	Stop transmitting a packet
PHY-RX-START	Prepare to receive a packet
PHY-RX-END	Exit receive mode
PHY-CCA-START	Begin CCA
PHY-CCA-END	End CCA

Table 3.20 PLME SAP services

Service	Description
PLME-GET	Obtain information from the PHY MIB
PLME-SET	Set information into the PHY MIB
PLME-RESET	Reset the PHY
PLME-RANGING-TIMER-START	Start the ranging timer
PLME-RANGING-TIMER-STOP	Stop the ranging timer

3.8.1 Management Services

The PHY provides a small set of management services through the PLME SAP interface, as shown in Table 3.20. The management services generally work by setting variables in the Management Information Block (MIB). The MIB variables that the PHY is expected to provide are shown in Table 3.21. Of course, as in all the elements described in this section, the designer may opt to provide other or different elements as desired.

The PHY may read or write the values of the MIB using PLME-GET request/confirm primitives.

The MAC sublayer may reset the PHY to the power-on state by issuing a PLME-RESET.request. The PHY provides support to the MAC for ranging services by providing a clock that the MAC may start with a PLME-RANGING-TIMER-START request and stop with a PLME-RANGING-TIMER-STOP request. The precision and depth of the ranging clock are provided to the MAC sublayer in the MIB.

3.8.2 Transmit and Receive Services

The transmit and receive services provide the data path between two devices. The MAC controls the timing of when these services are invoked, just like it does any

Table 3.21 PHY MIB values

Name	Type	Valid Range	Description
pMaxFramePayloadSize	Read only	0–4095	Payload size supported by the PHY
pPowerState	Read/write	SLEEP, STANDBY, READY	Power state of the PHY
pCCAStatus	Read only	BUSY, CLEAR	State of the wireless medium
pRCLKOptions	Read only	Refer to WiMedia specification	Ranging support capabilities
pRCLKTolerance	Read only	0–255	PHY ranging timer accuracy (ppm)
pRangingTimer	Read only	32-bit integer	Value of the ranging timer

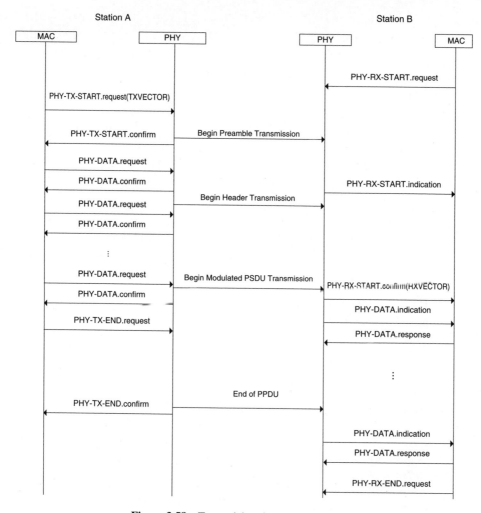

Figure 3.58   Transmit/receive sequence charts

PHY service. Figure 3.58 shows the sequence of messages for the transmit and the receive services.

To command the PHY to transmit, the MAC issues a PHY-TX-START.request and provides the necessary information: the TXVECTOR already described in Table 3.12.[4] The PHY will immediately begin transmitting a preamble and reply with a PHY-TX-START.confirm. The MAC may then transfer the payload and FCS using PHY-DATA.request and PHY-DATA.confirm primitives. When the MAC has

---

[4] The standard does not specify how the TXVECTOR is to be transferred. It is, of course, up to the implementer, but PHY-DATA.request/confirm primitives could be used for this purpose.

transferred all of the data, it issues a PHY-TX-END.request, and when the PHY has completed transmission of the PPDU it replies with a PHY-TX-END.confirm. Note that there is no mechanism to abort a transmission midstream, short of resetting the PHY. It is generally a bad idea to send incomplete packets in a networked environment.

The receive service is similar to transmit. The MAC issues a PHY-RX-START.request and the PHY enters receive mode. The PHY simply waits until it observes the presence of a PPDU by detecting a preamble. Of course, the MAC may abort reception any time by issuing a PHY-RX-END.request primitive.

When the PHY detects a packet, it will issue a PHY-RX-START.indication to the MAC. When the header has been received, the PHY issues a PHY-RX-START.confirm to the MAC and transfers the RXVECTOR to the MAC using PHY-DATA indication/response primitives. The RXVECTOR contains the header elements plus some reliability information, as shown in Table 3.22. The PHY next transmits the payload and FCS using PHY-DATA indication/response primitives. When the last datum has been transferred (the MAC will know because it has the length of the packet from the RXVECTOR), the MAC will issue a PHY-RX-END request and the PHY will be ready for subsequent command primitives.

Table 3.22 RXVECTOR elements

Parameter name	Description
Length	The number of octets in the payload. This length does *not* include the FCS
Data rate	This is the rate at which the PSDU portion of the packet is to be transmitted, and is one of the elements from Table 3.8
Burst mode	Informs the PHY that the next packet is a Burst-mode packet; in other words, it will follow the current packet with an MIFS
Preamble type	Informs the PHY that the next packet will have either a Burst mode or standard preamble
Scrambler seed	The final two bits of the scrambler seed, so that the receiver and transmitter scramblers may be synchronized
TX power	A number specifying the transmit power level
TFC, band group	Specifies the channel the PHY will be transmitting on. Note that only the LSB of the band group is specified, so the PHY will need this information from a different path
MAC header	10 bytes of MAC header information
Null tones	A bit field specifying nulled tones
Header error	Indicates whether the header was received correctly and the parameters were valid
RSSI	Receive signal strength indication
LQI	Link quality indicator

3.8.3 CCA Service

The PHY provides a CCA service to the MAC. This is simply an indication of the current state of the medium. If the PHY senses the presence of other transmissions then it will set the pCCAStatus variable in the MIB, and the MAC may read the MIB value as described above. The MAC will start the CCA service with a PHY-CCA-START request and stop it with a PHY-CCA-STOP request.

The WiMedia standard does not indicate a technique for performing CCA, but does state that the CCA must detect a busy channel with greater than 90 % probability within 5.625 μs when the signal at the receiver input is −80.8 dBm.

References

[1] Stockham, T.G., ' High speed convolution and correlation,' 1966 Sprint Joint Computer Conference, '*AFIPS Proceedings*,' vol. 28. American Federation of Information Processing Societies, Inc., Montvale, NJ, 1966, 229–233.

[2] Van Nec, R. and Prasad, R., ' *OFDM for Wireless Multimedia Communications*,' Artech House Publishers, Boston, MA, 2000.

[3] Schulze, H. and Luders, C., ' *Theory and Applications of OFDM and CDMA*,' John Wiley & Sons, Ltd, London, 2005.

[4] Proakis, J.G., ' *Digital Communications*,' McGraw-Hill, New York, 1995.

[5] Lin, S. and Costello Jr, D.J., ' *Error Control Coding: Fundamentals and Applications*,' Prentice-Hall, New York, 1983.

[6] Benedetto, S., Biglieri, E. and Castellenai, V., ' *Digital Transmission Theory*,' Prentice-Hall, New York, 1987.

[7] Molisch, A.F., Foerster, J.R. and Pendergrass, M., ' Channel models for ultrawideband personal area networks,' *IEEE Wireless Communications*, 10, 2003, 14–21.

[8] Di Benedetto, M.G. and Giancola, G., ' *Understanding Ultra Wide Band Radio Fundamentals*,' Prentice Hall PTR, Upper Saddle River, NJ, 2004.

[9] Khawza, A.I. and Heidari-Bateni, G., ' Improving two-way ranging precision with phase-offset measurements,' IEEE Global Telecommunications Conference, 2006. GLOBECOM '06, San Francisco, CA, 1–6.

4

Medium Access Control Sublayer

The Medium Access Control (MAC) sublayer is part of the Data Link layer as defined in the ISO/OSI-IEEE 802 reference model for a hierarchical communication architecture (see Figure 1.13).

The MAC sublayer provides medium/resource control through which upper layers (its clients) can efficiently and effectively communicate with their counterparts in other devices. In ECMA-368 [1], the MAC sublayer makes sure the users of the medium (the devices) do not interfere with each other even if they are using different clients/applications. It also allows for dynamic control of the communication parameters in order to address the varying wireless channel conditions that can cause packet data corruption or loss.

The MAC sublayer is designed for peer-to-peer, ad hoc networking. It offers guaranteed quality of service through the use of the TDMA technique for providing medium access. It is also capable of providing security (data encryption, message authentication, and replay attack protection), hibernation, aggregation and fragmentation, retransmission management (through a variety of acknowledgment policies), mobility management (through beacon management), fair reservation management, and ranging.

Obviously, to provide such capabilities in a robust fashion requires a sophisticated MAC protocol. The complexity is mostly due to the distributed nature of this MAC. Since no device is the master of the operation, there is no central authority, nor is there a central repository of information about all the neighbors in the network. Without such a central source of information or decision making, each device must maintain its own database of information about its neighbors and must follow relatively complicated protocols to make sure the network of devices operate in coordination with each other. To exacerbate the problem, the MAC is expected to allow for mobility of devices, which causes dynamic changes

WiMedia UWB: Technology of Choice for Wireless USB and Bluetooth   Ghobad Heidari
© 2008 John Wiley & Sons, Ltd

in this distributed network, including devices joining or leaving the neighborhood without notice.

Nevertheless, the complexity of the WiMedia MAC protocol is certainly worth the throughput, QoS, mobility, security, and network reliability that it affords. The MAC enables a very capable distributed ad hoc network that does not require a master of ceremony (such as a personal computer Host in USB or an Access Point in IEEE 802.11). This means that, for a particular link between two devices, a manufacturer does not have to build two different types of device (a master device and a slave device). Plus, the generic devices are able to form networks among themselves and devices can come and go without any user involvement or planning. An added benefit of not having such a master is the independence of the network from any one node or device in the network. Should any node in the network fail to operate, the rest of the network can still function properly, whereas the entire network is brought down if the controlling device in a master–slave network dies or severely malfunctions.

The MAC protocol also consists of convergence policies that allow diverse devices with different MAC clients (e.g. Wireless USB, Bluetooth 3.0, WLP) to share the channel without interfering with each other. The policies make sure that even though the distinct MAC clients do not know about each other, they all coexist peacefully and equitably over the same medium.

The MAC sublayer, as defined in ECMA-368, acts as the brain of the WiMedia PHY, controlling the PHY as a master does. The MAC receives services from the PHY and provides services to its own upper layer client through what are called Service Access Points (SAPs), as shown in Figure 1.14. In general, the SAPs may or may not be standardized. Within WiMedia, the PHY SAP is standardized (but is not mandatory to implement). The reason for this is to allow PHY semiconductor developers to build to a standard interface to interoperate with any MAC implementation. The MAC SAP has, to some extent, been defined in the MAC specification; however, it is by no means considered complete or mandatory. It is only offered as an informative part of the specification.

The PHY has very little autonomy when it comes to its operation. Most of the time, the MAC dictates whether the PHY will receive, transmit, do channel assessment, use a certain hopping pattern (channel), go to sleep, etc. Even the exact time of TX or RX is controlled very tightly by MAC on a packet-by-packet basis through an adequately defined MAC-PHY Interface (MPI) (defined in ECMA-369 [2]). The few places where PHY has its own autonomy are the following:

- During packet detection, PHY is able to look for packets and only notify MAC if it receives one.

- In Burst mode, due to the short duration of the Minimum Inter-Frame Spacing (MIFS) of this mode (1.875 μs), the PHY takes over the packet timings after the first packet.

The MAC assumes the following services from the PHY:

- frame transmission (TX);
- frame reception (RX);
- burst mode TX/RX (if supported);
- PLCP header error check;
- clear channel assessment;
- range measurement timestamps (if supported).

Unfortunately, the MAC specification is written in a very cryptic fashion. This is nothing new for a standard specification; however, the complexity of this MAC protocol aggravates this problem. It is the goal of this chapter to untangle the specification in an easy-to-follow and intuitive way, adding justifications and practical insights along the way. This chapter will build upon the PHY chapter (Chapter 3) in developing an understanding of the basic mechanisms of operation and control of the WiMedia UWB transceiver.

We will first start with a summary of the MAC features (Section 4.1). This will allow the reader to become familiar with all the capabilities and offerings of the WiMedia MAC without getting bogged down in the details. Then we will delve into the important facets of the MAC protocol and explore each in sufficient detail while keeping the big picture in clear view. The objective is to make the reader aware of and remove any confusion about how the different parts of the protocol work together to provide the desired level of control.

Throughout this chapter, we will adhere to the terminology of Section 1.7.

4.1 Feature Summary

4.1.1 Peer-to-peer

The WiMedia MAC is a purely peer-to-peer protocol, which refers to the fact that WiMedia MAC considers every device to be truly a peer of any other in the network. There is no hierarchy or master–slave relationship among devices. All devices are considered equal in their rights to the shared channel relative to each other. This is in contrast to other MAC protocols such as:

- USB or CW-USB, where one device (the Host) is the master and controls the entire network of other devices connected to it in a star formation;

- Bluetooth, where, although all devices are born equal, in any given piconet (grouping of devices), one of them will take on the role of the master;
- IEEE 802.11, when Access Points are used.

4.1.2 Ad Hoc

The MAC is designed to enable ad hoc networks. The network of WiMedia devices is expected to be created and modified without any preplanning. Devices are expected to find, join, or leave a network of other devices without any prior notice. Even disjoint groups of devices are expected to merge when they come close to each other. By the same token, a network of devices may easily split into smaller groups as devices move away from each other.

4.1.3 Two-hop Networking

Two-hop networking refers to the fact that this MAC protocol is designed to be aware of devices up to two hops away (neighbor of the neighbor). This is important because it allows the protocol to be resilient to the infamous hidden-node problem (see Section 4.1.4). By taking into account information about its one-hop and two-hop neighbors, a WiMedia device is able to avoid channel scheduling conflicts among itself, its neighbors, and its neighbors' neighbors.

4.1.4 Prioritized Contention Access (PCA)

Carrier Sense Multiple Access (CSMA) is the more traditional method in MAC protocols (e.g. IEEE 802.11) of allowing devices to access the medium. This method does not require a priori channel access negotiations. Instead, each device tries to access the channel when it finds an opportunity. Of course, it is expected that there will be collisions between devices whose transmissions overlap in time, rendering such received packets useless. As such, by adding Collision Avoidance to CSMA (CSMA/CA), each device is required to not only check (sense) the channel before transmission but also incorporate a random delay into its subsequent attempts at channel sensing. The drawback of this method is the wastage of throughput and energy by each device to keep sensing the channel until a transmit opportunity arises. Moreover, due to the unpredictability of channel access, guaranteed quality of service (QoS) cannot be provided.

Note that two devices may sense the channel at the same time and simultaneously conclude that the channel is not busy. Their subsequent transmissions may result in collisions. Thus, even with channel sensing and collision avoidance there is still the potential for collisions. Plus, CSMA is sensitive to the 'hidden-node' problem that, if one device finds the channel unused, it does not mean that its neighbor finds it the same way. That is, a device may receive simultaneous communication attempts from two of its neighbors who do not hear each other.

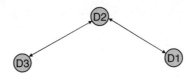

Figure 4.1 A simple hidden-node scenario

See Figure 4.1 for a simple scenario depicting the hidden-node problem. In this scenario, nodes D1 and D3, both within the range of node D2, cannot hear each other. In a traditional CSMA/CA protocol, nodes D1 and D3 may simultaneously transmit packets to node D2 since they are not able to sense each other. Their transmissions to D2 may result in collisions. Consequently, collisions may occur even if carrier sensing indicates no channel activity. These collisions would usually necessitate retransmissions, which further reduce throughput and increase power consumption.

The WiMedia MAC offers CSMA/CA and calls it PCA. The PCA operation is not mandatory to implement at the MAC level, but it may be a service required by its client (e.g. WLP).

4.1.5 Distributed Reservation Protocol (DRP)

TDMA is an old technique of providing shared access to multiple users of a resource (in this case an RF channel). In the WiMedia MAC, this method is termed DRP. Each device is able to announce its current channel usage time slots or negotiate its desired time slots with other devices. Compared with PCA, DRP has the benefit that each device can be aware ahead of time when it can transmit or receive without any need for time- and power-consuming channel sensing. The drawback is the need for coordination between devices of their transmit/receive slots. As we will see, this coordination is done through the use of Beacons sent by each user during prespecified time slots.

The DRP capability of the WiMedia MAC promises to provide throughput efficiencies (ratio of throughput at MAC level to data rate at PHY level, or T/R) far exceeding those of a WLAN. In IEEE 802.11a/b/g, a 50% T/R is about all that could optimistically be expected from its MAC protocol. For example, for a 54 Mbps IEEE 802.11a/g, practical throughputs are at most about 25 Mbps. However, for ECMA-368, the MAC protocol is expected to have a T/R ratio of 70–95%, depending on the channel condition. Table 4.1 lists the expected throughputs and T/R ratios for different link speeds. The values are given for both packet error rates of 0% and 8%. Note that the higher the packet error rate, the more retransmissions are necessary and, as a result, the lower the throughput.

Table 4.1 WiMedia MAC throughput and T/R projections

Data rate R (Mbps)	Secure payload Block size = 16 PER = 0%		Secure payload Block size = 16 PER = 8%	
	Throughput T (Mbps)	T/R (%)	Throughput T (Mbps)	T/R (%)
53.3	50.4	95	46.4	87
80	74.6	93	68.6	86
106.7	98.4	92	90.6	85
160	143.2	89	131.7	82
200	175.9	88	161.8	81
320	266.2	83	244.9	77
400	321.4	80	295.7	74
480	369.9	77	340.3	71

4.1.6 Mobility

WiMedia MAC is designed with mobility in mind. In this context, mobility is not the same as what one would expect from a mobile/cellular communication. Far from it; in this MAC, mobility simply means that it is possible for devices to move. Thus, neighbors may come and go. This is quite a challenge for the MAC, in the sense that the protocol must be sufficiently robust to withstand unexpected changes in the network topology due to mobility. This is in addition to the fact that the MAC has to be stable under wireless conditions (unexpected deterioration of links among devices). Hence, much attention has been paid to maintaining a level of stability that can withstand the comings and goings of devices in the ad hoc network.

4.1.7 Beaconing

Beaconing is a techno-jargon referring to the act of sending Beacons (Beacon packets). Beacons are special packets that carry most of the control information necessary for the distributed network to exist and to be stable. They may also carry information about channel reservations that each device makes. MAC clients can also send their layer's control information in the Beacons in the form of Information Elements (IEs). Each device is expected to send Beacons during predefined time slots. These time slots are chosen by each device through a process that tries to make sure no two devices choose the same Beacon Slot.

Beacon Slots are specially designated time slots in which each device announces its status, reservation request, reservation confirmation, etc. in the form of Beacon packets to all other devices. The act of sending Beacons means that a device has joined or created a neigborhood of devices, termed a Beacon Group. If the device does not find other devices to join, then it can start its own Beacon Group.

A Beacon Group, a device-dependent list, is the collection of the one-hop neighbors of a device. The Beacon Group of one device is not necessarily the same as those of other devices in its Beacon Group. It is quite possible that the set of neighbors of a device is different from that of its neighbor. Thus, it is very important not to make the mistake of assuming that the Beacon Groups of two neighboring devices are identical.

Of course, coordination must be maintained among devices to make sure the Beacons sent by all members of all overlapping Beacon Groups are noninterfering. As such, there are strict rules for a device on how to determine what Beacon Slot to use and how and when to change the Beacon Slot.

Different device neighborhoods merge together as they move within range of each other. Members of an extended (multi-hop) neighborhood are identified by their Beacon Period Start Time (BPST). They are expected to be on the lookout for 'Alien Beacons,' i.e. Beacons from other devices/neighborhoods that have a different BPST. When such Alien Beacons are found, merging activities may start to synchronize the BPSTs.

4.1.8 Band Groups and Time-Frequency Channels (TFC)

As discussed in the PHY chapter (Sections 3.3.1 and 3.3.2), each physical RF channel in WiMedia standards refers to an RF *band group* consisting of three RF bands of 528 MHz each. In the 3.1–10.6 GHz spectrum allocated to UWB operation, there are four such physical channels (band groups) plus a fifth band group made up of only two bands of 528 MHz (see Figure 4.2). In the first version of the standard, the first band group (3.168–4.752 GHz) was mandatory and the rest were optional to implement. In the latest revision [1], however, the mandatory requirement on Band Group 1 was lifted (since this band group is not universally available). Now, the choice of band group(s) to implement in PHY is left to implementation (as long as at least one band group is there). Additionally, the

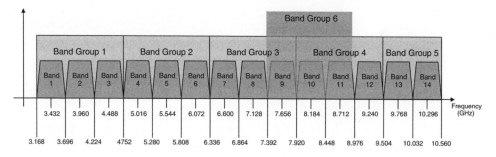

Figure 4.2   WiMedia (MB-OFDM) band groups

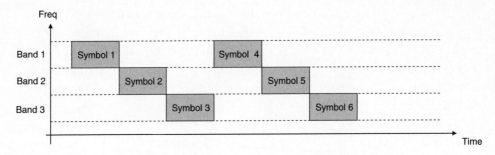

Figure 4.3 Hopping pattern for TFC 1 of Band Group 1

new revision of the specification debuted Band Group 6, which is an overlay band group, as shown Figure 4.2, consisting of Bands 9, 10, and 11. As explained in Section 2.6, from a regulatory perspective, this band group is more globally accepted.

In addition to the physical channels, the PHY and MAC specifications also call for several logical channels for each band group, all of which are mandatory to implement. The logical channels differ in their hopping patterns. These logical channels are referred to as TFCs. For Band Groups 1–4 and 6, the first four TFCs (TFC 1, 2, 3, 4) hop over the three bands in the band group with a variety of hopping patterns. The next three TFCs (TFC 5, 6, 7) do not hop at all, and the last three TFCs (TFC 8, 9, 10) hop over only two bands. For Band Group 5, which contains only two bands, TFCs are defined to either hop over the two bands or not hop at all. Figure 4.3 illustrates the hopping pattern of TFC 1. The details of the TFCs are given in Section 3.3.2. From the MAC standard perspective, the choice of the band group and TFC is made either at the client level or the DME. The MAC simply relays to the PHY which band group and TFC to operate in.

4.1.9 Acknowledgment Policies

WiMedia MAC allows for three modes of ARQ:

- No-ACK;
- Immediate ACK (Imm-ACK);
- Block ACK (B-ACK).

The first one is obvious: no acknowledgment is sent in response to received frames. Immediate ACK, on the other hand, requires the receiving device to send an acknowledgment frame back to the transmitting device for every received frame from that transmitter. The ACK is sent immediately after each received frame.

Note that here are no No ACK (or NAK) frames to indicate nonreceipt. That means the transmitter will have to infer nonreceipt based on the lack of ACK frames at the expected time. That is why there are very strict timing requirements for ACK frames to be sent by the receiver relative to the end of the last packet received. Of course, it is completely up to the transmitter whether or not to retransmit a frame that was not received. This is an implementation and/or MAC client choice and usually depends on factors such as type of data, application latency requirements, etc.

Obviously, sending an ACK frame after every received packet could waste quite a bit of channel resources. Consequently, for a more efficient channel usage, the Block ACK mode is offered in the MAC specification. After an initial handshaking with the receiver, the transmitter may transmit a block of packets and the receiver may send a B-ACK frame in response. The B-ACK frame contains the acknowledgments for each of the frames properly received in the previous block of transmitted frames. Any frame that was not acknowledged may be retransmitted among the next block of frames.

The drawback of Block ACK is that it requires more memory in the MAC sublayer to buffer frames while retransmissions are in progress. On the transmitter side, the MAC needs to keep the block of transmitted frames in the memory until they are all acknowledged by the receiver. If any one of them is not acknowledged, then the transmitter may have to retransmit it. Thus, the buffer cannot be released until all frames within a block are acknowledged (or until the frames become stale). Similarly, on the receive side, the MAC is expected to send the frames to its client (upper layers) in the order they were generated. As such, if a frame in a block is not received properly, then the MAC has to wait for its retransmission before all the subsequently received frames may be passed on to the MAC client. The size of the block in Block ACK is determined in the handshaking process between the two devices. In practice, the size depends on the amount of memory buffer the devices have.

4.1.10 Reservation Policies

The MAC policies allow the upper layer protocols to coexist without having to coordinate directly with each other. Referring to Figure 4.4, the transport and network protocols such as WUSB, Bluetooth, IP, and/or IEEE 1394[1] may be using the same UWB platform (PHY and MAC) without knowing about each other. Without a set

---

[1] The 1394 Trade Association is no longer pursuing WiMedia UWB as a platform at this moment.

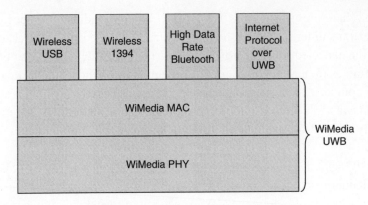

Figure 4.4   WiMedia UWB as a platform

of 'good neighbor' policies for these networks to operate by, it is likely that each of them will try to use the channel resources to its own benefit. Thus, it is up to the MAC sublayer to set policies for *all* devices to follow. These policies will ensure that no set of devices (within each network) will dominate the channel resources to the point that other protocol networks may become unfairly disadvantaged.

These MAC policies were originally envisaged to be actually a separate layer between MAC and its clients. However, they were eventually merged into the MAC specification as one of the annexes. The policies define the restrictions to be applied to reservations made by each device. For example, they limit the total amount of channel time, the amount of contiguous channel time, and the amount of periodic channel time that may be reserved by any one device. These policies are very important from a compliance perspective and will be discussed in detail in Section 4.11.

4.1.11 Security

Security is an optional feature of the MAC specification. However, this feature may be mandatory by upper layers. For example, in CW-USB, the security feature of the MAC is mandatory. Hence, the decision whether or not to develop this feature must take into account the requirements of the MAC clients or the applications.

The security features of the MAC sublayer include message authentication and data encryption, but not device association. There are very important distinctions among authentication, encryption, and association.

Association is the act of creating an initial relationship between two devices so that future data communication can be trusted to be between the intended devices. This initial trusted relationship is expected to be performed through an 'out-of-band' process. This means that either the messages transferred between the intended devices do not go over the same medium as the wireless channel used for data communication between those devices or, if they do,[2] that there is user intervention (which is the out-of-band part of the process) to prevent the man-in-the-middle attacks. For example, the out-of-band connection could be through a one-time wired connection or through human interaction. Other methods such as Near Field Communication (NFC) have also been suggested for a more user-friendly association process. However, they all do the job of confidently introducing the devices to each other so that no mistake or fraud in identifying the intended devices would take place. As a by-product of the association process, the devices exchange master keys to be used for secure communication among them.

Note that the MAC sublayer does not provide any association model for the device. This is left to the upper layers. The MAC assumes that, somehow, a shared secret key is established between the devices.

Authentication refers to the process of ensuring that a packet received from a device is trustworthy. This is to prevent the reception of packets from a device that tries to take on the identity of another: the so-called man-in-the-middle attack The MAC sublayer specifies a methodology for computing a Message Integrity Code (MIC) based on the Advanced Encryption Standard (AES) algorithm. Session keys are generated from the master keys and are securely exchanged among the devices. This allows for each device to authenticate each message (frame) that it receives from its secure neighbors. Note that authentication only protects the intended devices from receiving fraudulent data frames. It does not prevent the unauthorized reception of the frames. That is the job of encryption.

Encryption at the MAC level is done through the same AES engine. All or part of the payload may be encrypted. Partial encryption is useful when the upper layer protocols require that all neighbors hear certain parts of the message (e.g. the header) while keeping other parts confidential. This is done using an encryption offset field in the frame header that tells the receiving devices at what point in the secure payload the encryption starts. Whether encrypted or not, the frame authentication is preserved using the MIC.

4.1.12 Hibernation

One of the key requirements of the WiMedia UWB is low power operation. As such, a hibernation mechanism is provided, whereby a device that has nothing to send or receive for a while may refrain from transmitting and receiving beacons for a number of superframes. This allows the device to shut down most of its power-consuming circuitry and potentially 'go to sleep' periodically. The device

---

[2] The Diffie–Hellman protocol is a method of establishing a shared secret key between two parties over an insecure medium, but it is susceptible to the man-in-the-middle attack. See Section 5.17.

announces its intention to go into hibernation (and its hibernation duration) before doing so. The device and its neighbors maintain its beacon slot in the beacon group. Any device wishing to communicate with a hibernating device will then make sure to wait for that device to wake up from hibernation.

Thus, a device that is mostly in a monitoring mode can conserve energy while still being considered as part of the neighborhood (the Beacon Group).

4.1.13 Fragmentation, Aggregation

In different circumstances, the MAC sublayer may decide to aggregate multiple MSDUs into a single MAC frame, or conversely to fragment an MSDU into multiple smaller MAC frames (MPDUs). The reasons why the MAC sublayer may do this can vary. For example, if the MSDUs are relatively small, then the MAC sublayer can save channel resources and increase throughput by combining them into larger frames, provided that the channel conditions allow for large frame transmissions. On the other hand, if the channel conditions are not suitable for large frame sizes, then the MAC may break a large MSDU into fragments to optimize the QoS.[3]

4.1.14 Ranging

The fact that UWB signals occupy a very wide frequency spectrum makes them quite useful for ranging purposes. The ultra-wide bandwidth means that, in the time domain, the signal pulses are very short, providing for a very high time resolution to use for accurate ranging.

The WiMedia PHY uses timer (counter) values at the start of the channel estimation sequence of each packet transmitted and received in order to measure the round-trip time of flight between a transmitter and a receiver. The MAC sublayer supports this interaction by providing specific frame formats to exchange these data between the two devices. For improved accuracy, multiple two-way measurements may be made. This then allows for a distance calculation based on the average time durations measured.

4.2 Superframes and Timeslots

Within a band group, and given a certain hopping pattern (TFC), medium access sharing is provided through a TDMA scheme. Within this TDMA protocol, four time resolutions are defined: Superframe, MAC Access Slot (MAS), Beacon Slot, and Backoff Slot. Figure 4.5 shows the first three relative to each other.

---

[3] In general, larger frames have a higher probability of packet error than smaller ones.

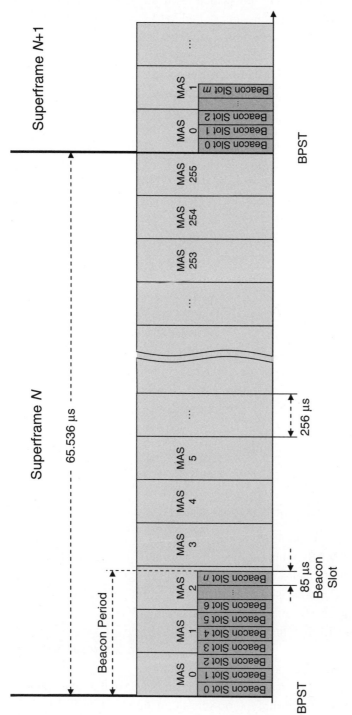

Figure 4.5 Superframe time slots

Superframe is a time duration which always starts with a BPST (see Section 4.8), and contains data transfer slots, called MASs. There are exactly 256 MASs in each superframe. Each MAS is 256 µs in duration, making the superframe exactly 65 536 µs long.

In addition to the MASs, there are Beacon Slots defined within a superframe. At 85 µs in duration, they are designated to contain Beacon packets transmitted by each device for the purpose of coordination with other devices. Beacon Slots overlay on top of MASs, starting from the beginning of a superframe (BPST). The first two Beacon Slots are designated for special use and are named Signaling Slots.

Since Beacon Slots are slightly shorter than one-third of an MAS and they are defined to be contiguous, the beginning of a Beacon Slot does not match with the beginning of any MAS, except for the very first Beacon Slot. This can be seen in Figure 4.5, where the Beacon Slots are misaligned relative to the MASs except at the beginning. (Note that, in every MAS, the offset between the last Beacon Slot in that MAS and the end of the MAS grows by 1 µs. That is, for instance, the 48th Beacon Slot in the 16th MAS ends within 16 µs of the end of that MAS.)

Each Beacon Slot may contain only one Beacon frame. Given all the different control information data units (IEs) defined for a Beacon, it is possible to construct a Beacon frame payload that exceeds in length the size of the Beacon Slot. Of course, it is not permitted to send such a Beacon. A Beacon packet, along with its SIFS and Guard Time must be completely contained within the Beacon Slot (see Section 4.4.1). If more IEs are necessary to transmit than fit into one Beacon Slot, then the device is expected to distribute the IEs over two or more Beacons (in two or more superframes). The likelihood of such a necessity is low, as the conditions under which a very large number of IEs need to be sent in a Beacon are rare.

The set of Beacon Slots of a superframe make up a Beacon Period (BP). All MASs that fall outside of the BP may be used for data, command, or control frame transfers. Conversely, no data/command/control frames may be transmitted during the BP.

Sometimes it is helpful to view the superframe in the form of a two-dimensional (2D) matrix of MASs. This is especially helpful in minimizing the number of bits required to identify the reserved MASs in an IE. In such cases, a 16×16 matrix of MASs is used to more easily refer to particular groupings of MASs, called zones. This 2D format of the superframe is depicted in Figure 4.6.

Note that as shown in Figure 4.6, the 256 MASs of a superframe are divided into 16 zones, each containing 16 MASs. Zone 0 contains the first 16 MASs of the superframe, and these are shown in the first column of Figure 4.6 from top to bottom. Similarly, Zone 1 is the group of the second 16 MASs of the superframe, and so on.

Backoff Slot is the fourth type of time slot defined within WiMedia MAC. It is 9 µs long and is used in connection with the PCA operation. It will be discussed further in Section 4.11.3.

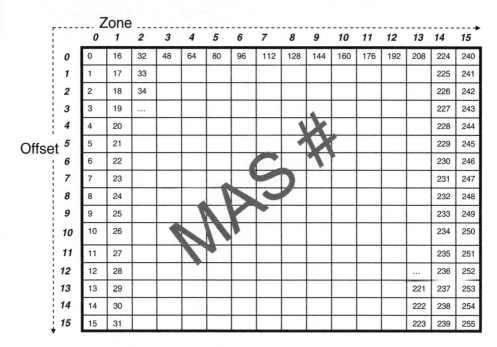

Figure 4.6 Two-dimensional superframe representation

4.3 Device Address

The MAC sublayer is identified by way of an EUI-48 (48-bit Extended Unique Identifier [3]). However, since each WiMedia UWB neighborhood is relatively limited in its maximum size, this 6-byte-addressing EUI-48 is wastefully too large for use in addressing the individual devices in a neighborhood. Instead, the Wi-Media MAC uses a 16-bit Device Address (DevAddr) that is usually randomly generated by each device as needed. This DevAddr represents the MAC temporarily. That is, every time a device generates a DevAddr, it may be a different one. DevAddr is associated with the EUI-48 for the duration it is used.

The address space used for DevAddr is divided into four groups.

- *Private*: 0x0000–0x00FF. Used by protocols such as CW-USB (see Chapter 5), in Private reservations, where the protocol does not use the WiMedia MAC addressing
- *Generated*: 0x0100–0xFEFF. All unicast DevAddrs that are not Private are Generated.
- *Multicast (McstAddr)*: 0xFF00–0xFFFE. Used for multicast addressing.
- *Broadcast (BcstAddr)*: 0xFFFF. A single DevAddr used for broadcast purposes (such as in Beacon frames).

A device randomly generates a Generated or Multicast DevAddr from the appropriate address range. DevAddr is expected to be unique over the two-hop neighborhood of the device. Therefore, each device has to check the generated DevAddr against those used by all other devices in its extended Beacon Group. The device does this by reviewing all DevAddrs mentioned in all BP Occupancy IEs (BPOIEs; see Section 4.4.1.1) received from its neighbors.

Even though the device makes sure its DevAddr is unique at the time of its generation, there is no guarantee that it will remain unique over the course of its usage. It is quite possible that two devices from two disjoint neighborhoods randomly select the same DevAddr. As long as the two neighborhoods do not come in to each other's range, there is no problem. Once they do, whether they merge or not, there is a DevAddr conflict that needs to be resolved. The MAC protocol provides for DevAddr conflict resolution.

A device can recognize that there is such a conflict if:

- it receives a frame with the same SrcAddr (Source Address) as its own DevAddr;
- it receives a BPOIE containing the same DevAddr but corresponding to a different device.

Once such recognition is made, the device is expected to generate a new DevAddr and start using it right away (in the next superframe).

4.4 Frame Formats

In this section we will quickly review the frame formats defined in the WiMedia MAC. The intent of this section, however, is not to define all fields of all frame types comprehensively. Such a treatment is already made in the standard and it would be duplicative (not to mention dry) to repeat that here. Instead, it would be a much more educational and more intuitive to revisit the different frame types and structures in the context of frame processing and general MAC protocol in the following sections. Hence, the subsequent sections will expand on the treatment of frame formats made in this section, as the need arises.

Recall that a PHY packet contains a PLCP preamble, PLCP Header (PHY header, MAC header and its channel encoding components), and PSDU, i.e. the payload plus its channel encoding and pad bits. A MAC frame is then defined as those components of the PHY packet that are of relevance to the MAC sublayer (including the PHY header). These are the elements of PLCP that are passed up to the MAC sublayer by the PHY. Of course, implementation depending, there may be other proprietary information that a particular PHY may pass to the MAC to help with the overall device performance. However, for this discussion, we will focus on standard elements of the MAC frame.

The general frame format of the MAC sub-layer can be seen in Figure 4.7. As shown, the MAC frame contains PHY header, MAC header, payload, and FCS. If the payload is secure, then the security fields (security header and MIC) are also included. The PHY header on its own contains the basic parameters of the PHY protocol, which are crucial to the MAC layer as well. These include the rate, length, scrambler, Burst mode, preamble type, TFC, and band group fields. (It also contains reserved bits for future expansion of the protocol.) Note that the PHY header is placed as early as possible in the PLCP Header. This is due to the timing requirements in the PHY, which needs to know the received data rate to prepare for the demodulation and decoding of the incoming payload before its arrival.

The MAC header contains frame control information. It tells the MAC sublayer what type of frame is being received, what ACK policy is being used, what the Destination and Source IDs are, what the sequence number of the frame is, etc. The MAC header fields are also shown in Figure 4.7.

Frame payload may or may not be secure. If insecure, it can range from 0 to 4095 bytes of user information. If secure, 20 of such bytes will be dedicated to security overhead. This is shown in Figure 4.7. Hence, the maximum secure user payload size is 4075 bytes.

Among the bit fields shown in Figure 4.7, the one which is least self-explanatory is Access Method. This is a 1-bit field referring to the type of reservation. The bit is set (set to 1) if the reservation type is such that it allows only the reservation owner to transmit (other than the necessary command/control responses sent by the reservation target) during the reservation block. As such, Hard DRP,[4] Private DRP, as well as certain circumstances in Soft DRP reservations[5] fall in this category and require Access Method to be set. On the other hand, all other types of medium access (including PCA[6] and Alien Beacon[7] reservations, as well as the nonqualifying Soft DRP users) must reset (set to 0) the Access Method bit. As an extension of this general rule, all command and control frames that are sent by a reservation target in response to received frames are expected to set their own Access Method bit to the same value as Access Method of the correspondingly received frame. Beacon frames are the exception to this general rule, in that their Access Method is reserved (set to 0). Table 4.2 summarizes the Access Method bit-setting rule described above.

There are six frame types defined in the MAC sublayer specification: Beacon, Control, Command, Data, and Aggregated. Besides these, there are three reserved

---

[4] DRP and its types are defined in Section 4.9.

[5] For example, when the owner of the Soft DRP reservation transmits without a back-off (see Section 4.9).

[6] PCA is defined in Section 4.11.3.

[7] See Section 4.8.4 for the definition.

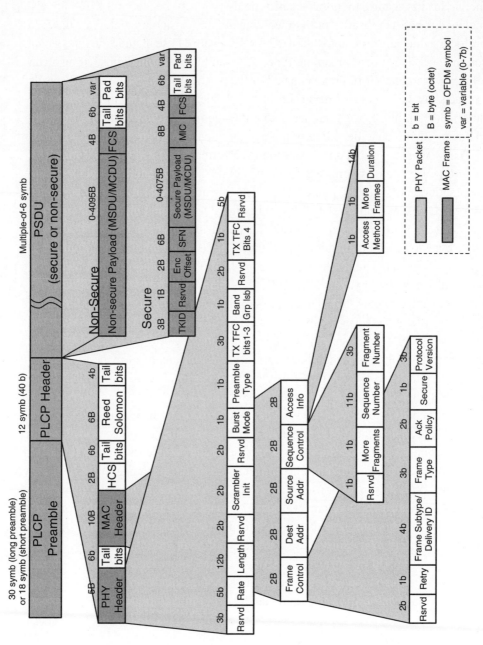

Figure 4.7 General MAC frame format

Table 4.2 Summary of Access Method bit-setting rule

Reservation type	Frame type/subtype	Access Method bit value	Notes
Hard DRP	All	1	This applies until a DRP reservation is released, inclusive of the UDA and UDR control frames. All command and control responses from reservation target shall copy the Access Method of the correspondingly received frame
Private DRP	All	1	This applies until a DRP reservation is released, inclusive of the UDA and UDR control frames. All command and control responses from reservation target shall copy the Access Method of the correspondingly received frame
Soft DRP	All	1	This applies only if reservation owner does not use back-off. All command and control responses from reservation target copy the Access Method of the correspondingly-received frame
PCA	All	0	
Alien Beacon	All	0	
	Beacon	Reserved	

frame formats for future expansion of the standard. Figure 4.7 depicts the general format of these frame types. For Beacon, Control, and Command frames, the payloads are used for signaling information. Although the Command and Control frames are available for transmission in any MAS during a superframe, it is expected that most signaling/command/control information is exchanged within Beacon frames during the BP.

4.4.1 Beacon Frames

Beacon frames are transmitted during the BP only. From a PHY perspective, the Beacon packet always contains a standard preamble and is sent at the lowest data rate (53.3 Mbps) to ensure the highest transmission reliability. Each Beacon packet is supposed to be transmitted at the beginning of a Beacon Slot and may not go over the duration of that slot. That is, unlike data frames that can straddle over multiple 'data' slots (MASs), Beacon frames are limited to what can be fit into a single Beacon Slot minus the usual Guard Time and SIFS duration (see Figure 4.8). Therefore, the maximum available duration for Beacon packet transmission is $85 - 12 - 10 = 63$ μs.

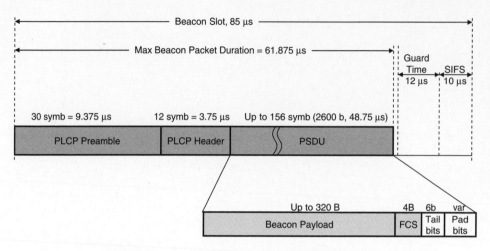

Figure 4.8   Beacon packet duration and payload capacity

Keeping in mind that the data rate for a Beacon frame is always at 53.3 Mbps (hence, 100 information bits per hop-frame; see Section 3.3.2), and subtracting the PLCP standard preamble (30 symbols = 9.375 μs) and the PLCP header (12 symbols = 3.75 μs), we end up with

Max. allowable Beacon PSDU duration = 63 − 9.375 − 3.75 = 49.875 μs

On the other hand, from a PHY perspective, a PSDU must always be a multiple of six symbols (a hop frame). Consequently, given that every symbol is 0.3125 μs:

$$\text{Max. number of symbols in a Beacon PSDU} = \left\lfloor \frac{49.875}{0.3125 \times 6} \right\rfloor \times 6 = 156$$

where $\lfloor x \rfloor$ is the largest integer smaller than x. As can be seen in Figure 4.8, this gives a PDSU of length an actual packet length of 61.875 μs. Given there are 100 bits of information in every 6 symbols at 53.3 Mbps, this further translates to 2600 bits of payload in Beacon PSDU, of which 32 go to the FCS, 6 go to tail bits, and 2 must go for padding (so that the remainder may be divisible by 8 to get an integer byte count). Thus, a maximum of 320 bytes of payload (MCDU) may be included in a Beacon frame.

Obviously, 320 bytes of payload is not very large. Currently, there are some corner cases in which a device may need to include in its Beacon more information than can fit in a single Beacon frame (more than 320 bytes). In the future, as the MAC clients such as CW-USB, WLP, or Bluetooth start using the MAC Beacons to include more and more application-specific IEs (ASIEs), the length limitation of the Beacon will be even more restrictive. The specification is silent as to what the device is supposed to do in such cases. For now, all such a device can do is to possibly split the optional IEs into two or more Beacon frames to be transmitted in two or more consecutive superframes. Care must be taken to have each Beacon frame still contain all the necessary fields and to distribute only the optional IEs over superframes. Of course, this method has its inherent drawback of delaying the transmission of control information over the air. The alternative is to send the extra IEs in Command frames during a superframe in one of the data (non-Beacon) time slots (MASs). Of course, aside from the fact that this alternative would take away from the data transmission capacity of the superframe, it would also require the device to spend more energy to accomplish it.[8]

For now, let us examine the format of a Beacon frame. The general format is, of course, the same as shown in Figure 4.7. Inserting the specific header and payload fields for the Beacon frame, we end up with the Beacon frame format of Figure 4.9. As seen in this figure, in a Beacon frame:

- the Destination Address is set to Broadcast Address;
- there are no frame fragments;
- most other header information are either set to 0 or are reserved.

The more interesting part of the Beacon frame is its payload. That is where most of the MAC-level control information is broadcast to the neighbors. The Beacon frame payload format is also shown in Figure 4.9. It basically contains Beacon parameters and IEs.

The Beacon parameters consist of Device Identifier (EUI-48, or 48-bit Extended Unique Identifier), Beacon Slot number (the slot number that the Beacon is supposedly occupying[9]), and the device control fields. The latter further contains, among other things, the bits to indicate if the Beacon is sent in the signaling slot, and whether or not the Beacon is movable. These concepts are described in Section 4.4.1.1.

The rest of the Beacon payload consists of a number of IEs (as many as necessary while maintaining the Beacon payload limit of 320 bytes). The general format of an IE is shown in Figure 4.9 and a list of the different types of permissible IEs is given in Table 4.3. In general, the IEs in a Beacon MCDU would have to be transmitted in the order of their Element ID. The only exception to this rule is that ASIEs can be inserted anywhere in the sequence of IEs but after IEs with an Element ID number less than 8.

---

[8] In the future revisions of ECMA-368, it is advisable for the specification to provide a methodology for assigning multiple Beacon Slots or longer Beacon lengths to a single device. There are a number of ways that this can be accomplished while staying backwards compatible.

[9] Note that even when the Beacon is sent in a signaling slot, its Beacon Slot number reflects its nonsignaling Beacon Slot number.

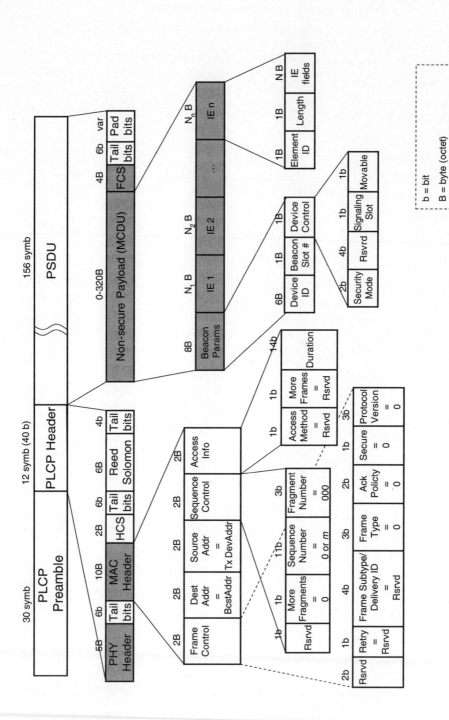

Figure 4.9 Beacon frame format

Table 4.3 Permissible IEs

Element ID	IE	Description
0	Traffic Indication Map IE (TIMIE)	Indicates device's intention to transmit data via PCA
1	BPOIE	Provides BP occupancy information collected by the device from its neighbors in the previous superframe
2	PCA Availability IE	Indicates the MASs in which a device is available/intending to receive/transmit via PCA
3–7	Reserved	Reserved
8	DRP Availability IE	Indicates the MAS's in which a device is available for new DRP reservations
9	DRP IE	Used to indicate or negotiate MAS reservations
10	Hibernation Mode IE	Provides hibernation information on a device that intends to sleep for one or more superframes
11	BP Switch IE	Indicates the device's intention to change its BPST
12	MAC Capabilities IE	Indicates the optional MAC capabilities the device supports, such as PCA, Hard DRP, Soft DRP, Block ACK, Explicit DRP Negotiation, Hibernation anchor, Probe, Link Feeback, Range measurement
13	PHY Capabilities IE	Indicates the optional PHY capabilities the device supports. These include TFI/FFI channels for Band Groups 1–5, and Reception data rates
14	Probe IE	Makes a request or responds to a request for one or more IEs from another device
15	Application-specific Probe IE	Makes a request for an ASIE from another device
16	Link Feedback IE	Makes a request to a source device to change its data rate and/or power control
17	Hibernation Anchor IE	Provides information on wake-up times of hibernating neighbors
18	Channel Change IE	Announces a device's intention to change its channel
19	Identification IE	Gives identifying information about the device, such as a name string, Vendor ID, etc.
20	Master Key Identifier (MKID) IE	Identifies some or all of the device's master keys
21	Relinquish Request IE	Used to preempt a device to relinquish some of its reserved MASs, usually as when the target device has reserved an excessive number of MASs. See Section 4.11 for definition of 'excessive'
22	Multicast Address Binding (MAB) IE	Provides mapping of each multicast EUI-48 to a 16-bit McstAddr
23	Tone-nulling IE	Provides tone-nulling information for use in DAA operation
24	Regulatory Domain IE	Provides information on the regulatory domain (country) for the device
25–249	Reserved	Reserved
250	WLP IE	This IE is reserved for WLP usage
251	WiMedia Platform Test IE	This IE is reserved for Platform Test usage
21–254	Reserved	Reserved
255	ASIE	Provided for Applications to define their own dedicated set of IEs

BPOIE

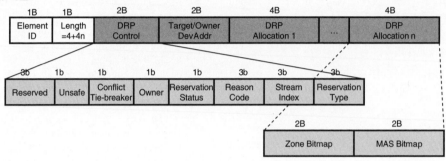

DRP IE

ASIE

Figure 4.10   Frame payload format of BPOIE, DRP IE, and ASIE

Some of the most important IEs are:

- BPOIE;
- DRP IE;
- ASIE.

The payload formats of these IEs are depicted in Figure 4.10. Besides BPOIE, DRP IE, and ASIE, there are many other important IEs that are not discussed in this section. The purpose of this section is to give an introductory view of IEs in general, not to describe each exhaustively. The reader is referred to the text of the ECMA-368 specification for further details on this topic. Let us now examine these three IEs more closely.

4.4.1.1 BPOIE

The BPOIE is the most important IE and one that has to be present in all Beacons. In this IE, a device gives information from its own perspective on the BP and its

Table 4.4   Beacon Slot Info Bitmap

Bit values	Occupied	Valid Frame	Movable
00	No	N/A	N/A
01	Yes	Yes	No
10	Yes	No	Yes
11	Yes	Yes	Yes

occupation by other devices. As can be seen in Figure 4.10, each device includes in this IE the BP length (in Beacon Slots), Beacon Slot Information Bitmap, and DevAddrs (Device Address) of neighboring devices from which the device received a Beacon in the previous superframe. The bitmap consists of 2 bits for each Beacon Slot in the BP indicating whether or not:

- the Beacon Slot is occupied;
- a valid Beacon frame (error-free header and payload) was received;[10]
- the Beacon is movable.

See Table 4.4 for the bit-mapping values of the Beacon Slot Info Bitmap and their interpretations.

DevAddr fields in the BPOIE represent the DevAddr of the occupied Beacon Slots in the Beacon Slot Info Bitmap, even if no valid frames were received. If a frame with invalid HCS (errored header) is received, the DevAddr of the frame will not be known. In this case, the BcstAddr (Broadcast Address) replaces the DevAddr field in the BPOIE. As we will see later in the hibernation discussion, a device that is in hibernation mode can reserve its Beacon Slot through a neighbor that acts as the *anchor*. In this case, the BPOIE contains the DevAddr of the hibernating neighbor and marks that Beacon Slot occupied and nonmovable. This way, when the hibernating device wakes up, it has its Beacon Slot to use right away without wasting time in re-establishing one.

4.4.1.2 DRP IE

DRP IE is arguably the second most important IE of Table 4.3. It provides a mechanism for devices to negotiate and announce their MAS reservations. It is the more efficient method of enabling the TDMA-based multiple access. The other method

---

[10] Note that the PHY may receive a valid PLCP preamble but fail to receive a correct header or payload. The MAC specification makes a special effort to identify such cases just in case there was a collision of beacons or otherwise a channel condition that caused the incorrect reception of the packet. In that case, the Beacon Slot is marked occupied so as to avoid any further collisions.

is through explicit DRP negotiations, which are command frames sent during regular MASs. Obviously, the explicit method takes up valuable superframe capacity that otherwise could be used for application data traffic.

The format of this IE is shown in Figure 4.10. It consists of:

- A DRP Control field, indicating:
  - whether the reservation is safe or unsafe (i.e. whether it is preemptable – see Section 4.11);
  - the Reservation Status (whether it is established or under negotiation/conflict);
  - if the device is the Owner of the reservation (the source device);
  - the Reservation Type (Alien BP, Hard, Soft, Private, or PCA);
  - the Stream Index (the data stream to use the reservation);
  - the Reason Code, when responding to a request for a reservation (Accepted, in Conflict, Pending, Denied, or Modified);
  - the Conflict Tie-breaker bit (which is set randomly by each reservation source/owner, and used consistently thereafter by all devices whenever referring to that reservation).
- Target/Owner DevAddr:
  - The owner device sets this field to the target's DevAddr (which could be a unicast or a multicast DevAddr), and the target device sets this field to the owner's DevAddr.
  - This field is Reserved if the Reservation Type is PCA or Alien BP.
- And one or more DRP Allocation fields, which are used to define the range of the MASs of interest in a concise manner. As seen in Figure 4.10, DRP Allocation fields contain two components:
  - Zone Bitmap, which is a 16-bit field, and each bit identifies one of the 16 zones in the superframe, as depicted in Figure 4.6. When a bit is set in this field, the corresponding zone (0–15) contains one or more MASs in the reservation.
  - MAS Bitmap, which is also a 16-bit field, and each bit identifies one of the 16 MASs in a zone, as depicted in Figure 4.6. When a bit is set in this field, the corresponding MASs (0–15) of all the active zones in the Zone Bitmap field are considered as part of the reservation.

To exemplify the use of DRP Allocation fields, we will consider a few different reservation scenarios.

Example 1

Let us code the reservation intended by the shaded squares in the MAS table of Figure 4.11. This is called a row reservation for obvious reasons. Note that this type of reservation is useful for applications that require periodic channel time

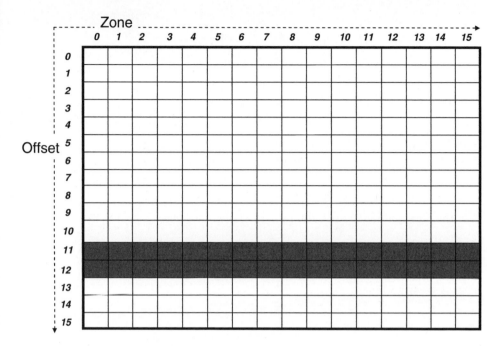

Figure 4.11 Example of a row reservation

allocations of short duration. Devices with small amounts of memory or applications requiring short service intervals may need this type of reservation to be able to transmit/receive their payload without the need for long delays or large data buffers. Also note that this type of reservation is not going to be very power efficient, as it requires the device to go into active mode frequently within a superframe.

If we show this reservation on the one-dimensional timeline of a superframe, Figure 4.12 will result.

To map this particular reservation using the DRP Allocation field format, we need to identify the zones of interest and the MASs of interest within each zone. In this simple example, all zones are of interest, since they all contain one or more MASs to be reserved. Within each zone, MASs 11 and 12 are reserved. Hence, the DRP Allocation field will be populated as shown in Figure 4.13.

Figure 4.12 One-dimensional version of MAS reservations of Figure 4.11

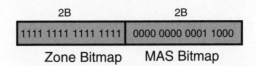

Figure 4.13   DRP Allocation field for the MAS reservations of Figure 4.12

Example 2

Now let us try a column reservation. Figure 4.14 shows such an example, where two columns of eight consecutive MASs within a superframe have been reserved. This kind of reservation is useful for video or bulk transfer applications where the buffer size and latency is not as much an issue as the power consumption and data transfer efficiency are. Figure 4.15 shows the one-dimensional view of this reservation.

Similar to Example 1 above, it is easy to map this particular reservation using the DRP Allocation field format. Here, zones of interest are only zones 4 and 12, and the MASs of interest within those zones are 0–7. Therefore, the DRP Allocation field will be populated as shown in Figure 4.16.

Example 3

Not all reservations can be fit into a single DRP Allocation field. That is why there are multiple such fields allowed in a DRP IE (as seen in Figure 4.10). For

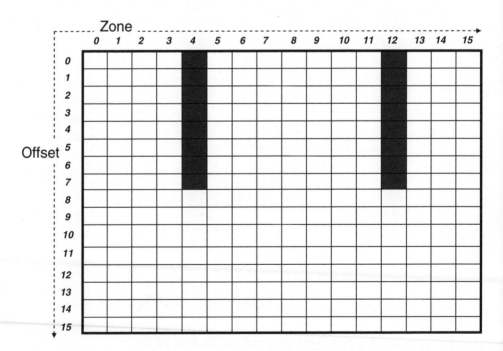

Figure 4.14   Example of a column reservation

Figure 4.15   One-dimensional view for reservation of Figure 4.14

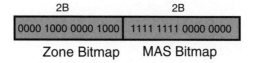

Zone Bitmap   MAS Bitmap

Figure 4.16   DRP Allocation field for the MAS reservations of Figure 4.14

instance, let us consider the irregular reservation of Figure 4.17. Here, we will need three DRP Allocation fields to represent the intended reservation. These fields are populated as depicted in Figure 4.18. The top field covers the row reservation in the middle of the MAS table of Figure 4.17. The middle field covers the column reservation in zone 8. And, the last field covers the remaining MASs to be reserved.

4.4.1.3 ASIE

An ASIE is another example of an IE that we will explore here. Its importance is not for the MAC layer, but for the layers above it. In fact, the ASIE is a place holder for all sorts of Information Elements that the clients of the MAC layer may want to define and use. An ASIE allows the MAC client to use the Beacon protocol and its frame format of the MAC to efficiently enable the client's protocol control requirements.

The format of this IE is shown in Figure 4.10. Besides the usual fields, the ASIE frame contains a Specifier ID field that is set to a 2-byte identification number representing a company or organization which is responsible for defining its own application-specific control frames, command frames, IEs, and Probe IEs. The idea is that each organization or company wishing to create its own custom MAC-level messaging to be sent among devices that subscribe to it may do so via the use of such application-specific constructs. This is especially useful for each PAL to be able to define its own IEs to be carried in the form of an ASIE among its devices. See Chapter 5 for more discussion on PALs. The updated list of Specifier IDs and the associated organizations are given at the Ecma website.[11]

The main field of the ASIE frame is the Application-specific Request Information, which contains messages in a format customized by the PAL/client/company. Note that this field may contain identifiers that allow a PAL, client, or organization to define a series of custom message/frame types using a single ASIE Specifier ID.

---

[11] Click on the link for Specifier ID Register at http://www.ecma-international.org/publications/standards/Ecma-368.htm.

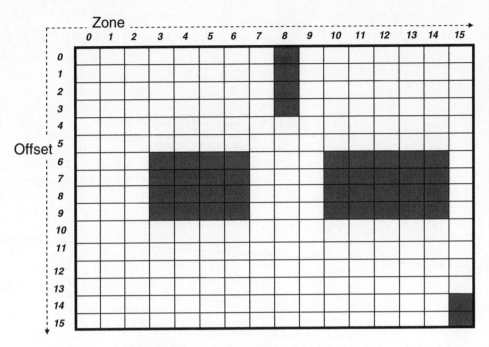

Figure 4.17   Example of an irregular reservation

4.4.2 Control Frames

Control frames are essentially used for the purpose of controlling traffic flow during regular MASs (the MASs that do not overlap the BP). They consist of special frame subtypes, control acknowledgments, PCA initiation or termination, as well as any application-specific controls. Table 4.5 shows the complete list and the description of control frame subtypes defined so far, with several of them reserved for future expansion of the list.

The general format of control frames is depicted in Figure 4.19. Note that control frames may be transmitted in secure or nonsecure mode, depending on the frame subtype. However, so far, there is only one subtype that is allowed for

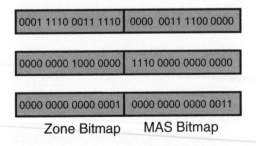

Figure 4.18   DRP Allocation field for the MAS reservations of Figure 4.17

Table 4.5 Control Frame subtypes

Value	Subtype	Description
0	Imm-ACK	Used for immediate acknowledgement of the correct reception of a frame
1	B-ACK	Used for acknowledgement of the correct reception of one or more frames
2	RTS	Used for initializing a PCA transmission. A device uses this control frame to declare to the receiver that it has data ready for transmission
3	CTS	Used by the recipient of the RTS control frame to declare the receiver's ability/readiness to receive a PCA-based data communication
4	UDA	Used by the owner of a DRP reservation block to declare the release of the remainder of that block for PCA usage by other devices
5	UDR	Used in response to a received UDA command by the target of a DRP reservation block to declare the release of the remainder of that block for PCA usage by other devices[a]
6–13	Reserved	Reserved
14	Application-Specific	Used by MAC clients or applications to define their own set of custom control frame subtypes
15	Reserved	Reserved

[a]Note that this is necessary because this MAC protocol is based on a two-hop topology. The neighbors of the DRP target device may not be neighbors of the DRP owner device. Thus, a UDA command is not sufficient to announce the release of a reservation block. A UDR is necessary to spread the word.

secure transmission, and that is the Application-Specific subtypes. That is, unless a MAC client or application has a custom control frame subtype, all MAC-defined subtypes are sent with nonsecure payloads.

Also note in Figure 4.19 that the Access Method for this frame type is not fixed (as it was in Beacon frames). As described in Table 4.2, the Access Method bit is dependent on the circumstance in which these control frames are used.

The most important of the Control Frame subtypes are Imm-ACK and B-ACK (subtypes 0 and 1). As previously described, there are three types of acknowledgments defined in this MAC protocol: No Acknowledgment, Immediate Acknowledgment, and Block Acknowledgment. The first does not require a control frame subtype, obviously. The latter two do.

The Imm-ACK has no payload, and its DestAddr is set to the SrcAddr of the frame to be acknowledged. B-ACK, on the other hand, contains a payload, as illustrated in Figure 4.20. This payload acknowledges individual frames that were correctly received in the last block transmission and gives information to the transmitter on the next block of frames to be transmitted. One such piece of information is the Buffer Size, which tells the transmitter how many bytes of total frame payloads the receiver is able to receive in the next block. Similarly, Frame Count limits

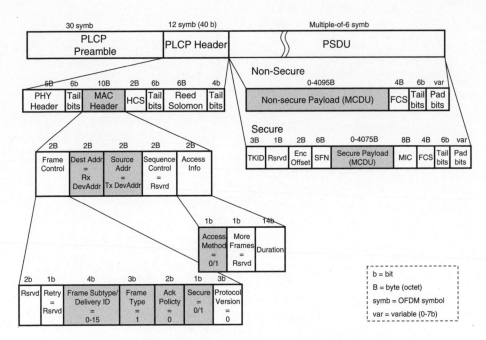

Figure 4.19 General format of Control Frames

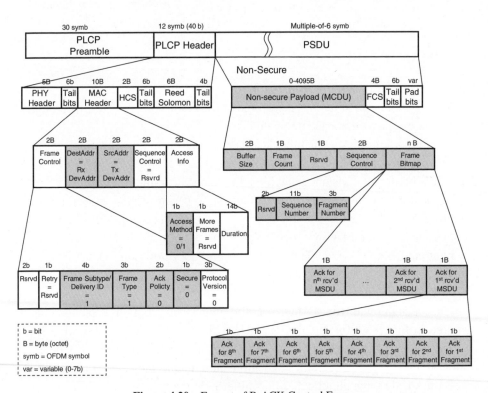

Figure 4.20 Format of B-ACK Control Frame

the transmitter on the maximum number of frames it can transmit in its next block transmission.

The acknowledgment in a B-ACK is done in a window format. The window is over the MSDUs that were received and have not been acknowledged yet. The starting point of this window is defined by the Sequence Number and Fragment Number fields in the B-ACK frame payload (see Figure 4.20). The Frame Bitmap field then acknowledges each individual frame that is correctly received since that starting point.

The Frame Bitmap field contains 1 byte for each MSDU that was supposed to have been received in the last block transmission. The least-significant byte of the Frame Bitmap field corresponds to the first MSDU in the sequence (the one with the sequence number given in the Sequence Control field). Each successively higher byte in the Frame Bitmap field corresponds to the next MSDU in the block received, as seen in Figure 4.20. Similarly, within each byte in the Frame Bitmap field, the bits are ordered as shown and they are set to 1 if the corresponding MSDU fragment was correctly received; otherwise they are reset (to 0).

Note that the B-ACK is the only case in which a NAK (acknowledgement of nonreceipt) is sent in this MAC protocol. In all other cases, if an expected frame is not correctly received, there is no indicative response from the receiver to the transmitter.

4.4.3 Command Frames

Command frames are used for the purpose of sending commands or requests from one device to another. Command frames come in a variety of subtypes as well. These are listed in Table 4.6. Some of these commands are there to provide duplicate capability that is normally available through the exchange of IEs in the Beacon frames. DRP Reservation Request/Response and Probe subtypes fall in this category. These allow for devices to have additional opportunities during a superframe to exchange such information. On the other hand, the rest of the subtypes are new types of command that are otherwise not possible to make in any other way, such as PTK and GTK, as well as Range Measurement.

The general format of command frames is shown in Figure 4.21. Note that most command frames may be transmitted in both secure and nonsecure modes.

4.4.4 Data Frames

Data frames are used for the exchange of the MAC client data among devices. It can be sent in unicast, multicast, or broadcast fashion. The general format of these frames is given in Figure 4.7.

Table 4.6   Command Frame subtypes

Value	Subtype	Description
0	DRP Reservation Request	Used to request the setup of or change in a DRP reservation
1	DRP Reservation Response	Used to respond to a DRP Reservation Request
2	Probe	Used to request the exchange IEs
3	Pair-wise Temporal Key (PTK)	Used, as part of the security protocol, in a four-way handshake between two devices to extract the PTK
4	Group Temporal Key (GTK)	Used, as part of the security protocol, to get/give GTK
5	Range Measurement	Used to exchange timing information related to range measurement
6–13	Reserved	Reserved
14	Application-specific	Used by MAC clients or applications to define their own set of custom command frame subtypes
15	Reserved	Reserved

4.4.5 Aggregated Data Frames

Aggregated Data frames are provided in this MAC to allow for a more efficient delivery of short MSDUs. When possible, the MAC can aggregate several MSDUs into an Aggregated Data frame to create the MPDU. The maximum size of an Aggregated Data frame is still 4095 bytes.

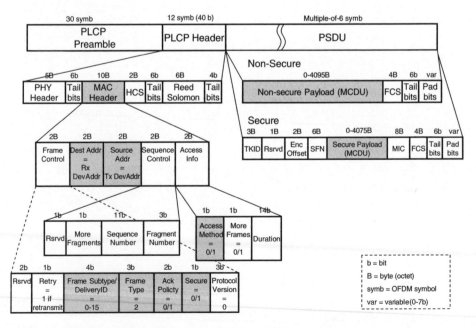

Figure 4.21   General format of Command Frames

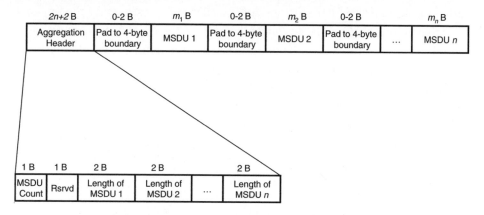

Figure 4.22   Payload format of Aggregated Data frame

When aggregating MSDUs, they must follow the frame payload structure given in Figure 4.22. Note that each MSDU is preceded with a padding of up to 2 bytes to ensure 4-byte boundaries are kept intact among them.

4.5 Frame Processing

In this section we will examine the MAC rules for processing the various MAC frames for transmission or upon reception. Depending on the frame type, these rules may relate to the MAC header fields as well as to the payloads.

4.5.1 Frame Reception and Transmission

Let us first consider the frame reception processing requirements at the MAC level. As seen in Figure 4.7, the MAC payload contains Frame Check Sequence (FCS) bits to detect whether the received frame payload contains one or more bit errors (FCS is same as CRC[12]). Note that FCS is normally generated (for transmit frames) and checked (on received frames) at the MAC level and not in the PHY.[13] On the other hand, as noted in Section 3.4.3, the HCS (another CRC), which is used to detect bit errors in the frame Header, is checked at the PHY layer of the receiver. The reason is that, once a packet arrives, the PHY must be able to tell if it has correct information about the incoming frame to receive it correctly. Of particular importance is the frame length. Without this information, the PHY never knows when to stop receiving the packet. Hence, the HCS provides the required confidence in the accuracy of the received frame Header. Once the PHY preamble

---

[12] Cyclic Redundancy Check.
[13] Of course, this is purely an implementation choice for the receiver, but given the timing requirements at the PHY level, it makes the most sense.

reception is complete, it starts passing the Header and subsequently the payload and FCS up to the MAC sublayer, where they can be stored. However, while passing up the Header bits, PHY is running its error correction (Reed–Solomon) and detection (HCS) algorithms on the Header. If the HCS bits do not match against what the PHY calculates, then it indicates Header error to the MAC and aborts the reception of the rest of the frame. It is normally up to the MAC sublayer at that point to restart the PHY to receive the next packet at the right time.

On the other hand, if no Header errors are detected, then PHY continues to receive the payload, along with FCS bits, and passes them all up to the MAC sublayer. Usually, the PHY does not have enough memory to keep the entire frame payload until the FCS bits arrive so that an error check can be made on the payload. Thus, the PHY passes the payload bytes to the MAC as they arrive. Once the whole frame is in, the MAC will run payload error checking (using FCS bits). If no error is detected, and the frame received was a Data or Aggregated Data frame, then the MAC constructs the MSDU to send to its client. (Other steps, such as de-aggregation or reassembly (defragmentation) may have to be taken before an MSDU is constructed from the payload received.) On the other hand, if payload error is detected, the MAC will discard the frame and may take remedial action, depending on the ACK policy (see Section 4.5.2). The MAC is not allowed to deliver to its client MSDUs out of order. Thus, even if retransmissions cause packets to be received out of order, the MAC must take care not to pass any out-of-order MSDUs to the upper layers.

Frame transmission rules require that all MSDUs be sequentially transmitted in the order they are received from the MAC client, as long as they are for the same device and the same delivery ID. If the destinations or delivery IDs are different, then the MAC sublayer may reorder those MSDUs for transmission. All bytes of an MSDU are to be transmitted in the same order as they are received. Similarly, each fragment of an MSDU must be transmitted in order. However, during a Block transmission, due to the fact that some MSDUs, or fragments thereof, may have to be retransmitted, these retransmissions may be out of order compared with the previously transmitted frames or fragments.

Since it is likely for a source device not to correctly receive an acknowledgment sent by the destination device for a frame it correctly received, a source may unknowingly send and the target device may receive a duplicate copy of a frame. In such cases, the Retry bit (see Figure 4.7) is set by the transmitter to indicate the retry attempt. By matching the Sequence Control, SrcAddr, DestAddr, and Delivery ID fields of such a frame against a previously received frame, the receiver can determine if they are duplicates. The target device has to discard the duplicate frame.

All Beacon frames are expected to be sent at the lowest PHY data rate, i.e. 53.3 Mbps. This is to ensure that Beacons (which receive no acknowledgments) get the best chance of being received by the neighboring devices. All other frame types may be sent at any PHY rate that the transmitting device desires, as long as the receiving device supports it. Recall that all devices have to support data rates 53.3, 106.7 and 200 Mbps in their PHYs.

4.5.2 Acknowledgments and Retransmissions

The Acknowledgment policy to be used in a transaction with another device is determined by the source device, subject to the capabilities of the target device. By filling the ACK field in the MAC header appropriately, in effect, the source device tells the destination device that it requires it to send the corresponding type of acknowledgment for all the frames it receives correctly. There are three types of Acknowledgment available in the MAC protocol:

- No-ACK (No Acknowledgment);
- Imm-ACK (Immediate Acknowledgment);
- B-ACK (Block Acknowledgment).

The first two are fairly obvious. No-ACK requires no acknowledgments from the destination device. This leads to the highest throughput, but at the risk of losing packets along the way. Streaming applications that cannot afford the delay in retransmission are good examples for use of No-ACK policy. Also, No-ACK is the only option when transmitting Broadcast or Multicast frames.

On the other hand, Imm-ACK requires an acknowledgment immediately after each packet is received. This leads to the lowest throughput but guarantees a timely response from the destination device, and it allows for immediate retransmission of failed transmissions. A compromise option is B-ACK, whereby the destination device will send an acknowledgment packet for a block of received packets. After each block of packets is received, the destination device sends a B-ACK to let the source device know which of those packets were received correctly and which were received in error or not received at all. In the next block of packets, the source device would then retransmit the packets that were not received or not received correctly.

In the case of B-ACK, the transmitter must first check with the receiver if it can handle block transmissions. B-ACK capability is optional in the MAC sublayer. A device that is capable of B-ACK advertises it in its Beacon. However, even if the destination device is indicating B-ACK capability in its Beacon, a source device must first check if the destination device is willing and capable of receiving block transmissions at that time. (Temporal buffer limitations could be preventing the device from receiving a block of frames.) To do this, the device sends a frame with B-ACK set to B-ACK Request. In response, the destination device must send a B-ACK frame with its payload containing the block size it is capable of receiving (see Figure 4.20). If the device intends to reject the B-ACK policy, then it sets the payload to zero.

All but the last frame in a block transfer is sent with Ack Policy field set to B-ACK (Figure 4.7). The last frame in the block sets this field to B-ACK Request. This serves as a B-ACK request to the target device for the next block of frames to be sent by the source. In particular, it indicates to the target device to send a B-ACK frame that acknowledges the previous block and to give the indication whether it will accept B-ACK transfer again and in what size. Thus, the target device has the option to stop or modify the parameters of the B-ACK transfer after each block is received. This option is very important, because at any given moment the receiver may run into a backlog at its interface to its client and, hence, may not be able to empty its buffers in time to receive the next block of frames.

Note that Imm-ACK and B-ACK acknowledgments are expected to be sent regardless of the security validity of the received frames.

The purpose of acknowledgments (whether Imm-ACK or B-ACK) is to allow the source device to retransmit any frames that either were not received or were received incorrectly. However, the choice of whether to retransmit or not is purely up to the source device. Depending on the QoS requirements of the application (among other factors), a source device may determine that a retransmission is not necessary or beneficial at times.

4.5.3 Frame Transaction and Inter-frame Spacing

The term Frame Transaction in the MAC sublayer refers to a single frame along with any associated Request To Send (RTS)/Clear To Send (CTS) exchange and any associated acknowledgment. As such, the act of a source transmitting a frame in a DRP mode and receiving an Imm-ACK frame back from the receiver is considered one Frame Transaction. Similarly, the act of sending one frame in a block transfer is a Frame Transaction, as long as the frame is not the last frame in the block. For the last frame in the block, one must consider the B-ACK response from the receiver as part of the Frame Transaction. Figure 4.23 illustrates some examples of Frame Transactions. Note that the examples are based on implicit DRP reservations, where the use of RTS/CTS exchange is not necessary.

Related to Frame Transactions is the inter-frame spacing. There are three types of inter-frame spacing defined in the MAC:

- SIFS = 10 μs;
- MIFS = 1.875 μs (six symbols long);
- Arbitration Inter-Frame Space (AIFS) = 19 μs, 28 μs, 46 μs, or 73 μs, depending on the user traffic priority: Voice, Video, Best Effort, or Background.

SIFS is the inter-frame spacing that is used to separate frames within a Frame Transaction. Also, it is the spacing required between any received frame and a subsequently transmitted frame. SIFS duration is fixed at 10 μs. It is the maximum time allowed for a given transmitter to turn itself around into a receiver, or vice versa. All transactions that require such turnaround time will need to include an SIFS period.

MIFS is the spacing used in the Burst mode of the PHY layer. In other words, Burst mode is achieved when the PHY transmits multiple frames in succession with MIFS spacing between them, increasing the overall throughput. See Section 3.5.1 for more detail on Burst mode. Note that the Burst mode can only happen with No-ACK or B-ACK.

AIFS is specifically related to the PCA mode of transmission, in which the device, after determining the channel is idle, must defer transmission for the duration of AIFS. Since each user priority, termed Access Category (AC), has its own AIFS value, the higher priority AC will get the first chance to use the medium in the PCA mode. See Section 4.10 for more information on the PCA mode of communication.

Figure 4.23 illustrates the SIFS and MIFS timings as well as the different ACK policies with a few examples.

4.5.4 Fragmentation and Aggregation

Recall that, for transmission, a MAC client passes an MSDU through the MAC SAP to the MAC sublayer. Since the MSDU sizes are usually decided by the MAC client without consultation with the MAC sublayer, there is a good chance that they are not optimal for transmission over the air. For instance, if the channel condition is poor, then it may be beneficial for the MAC sublayer to transmit using smaller packet sizes to ensure retransmissions require fewer bits per packet. On the other hand, if the channel condition is great, then aggregation of smaller MSDUs into larger packet sizes could improve overall throughput and reduce power consumption.

In addition to MSDUs, the MAC sublayer may generate MCDUs to communicate with a peer MAC sublayer on another device.

In general, the MAC may fragment or aggregate MSDUs/MCDUs to construct MPDUs, which are what are actually sent to the PHY for transmission. In WiMedia MAC, any MSDU/MCDU may be fragmented into up to eight MPDUs, each potentially with a different size (from 1 to 4095 bytes). The fragments are sequentially numbered, starting with zero, and the number is inserted in the Fragment Number field of the MPDU (see Figure 4.7). All fragments get the same Sequence Number.

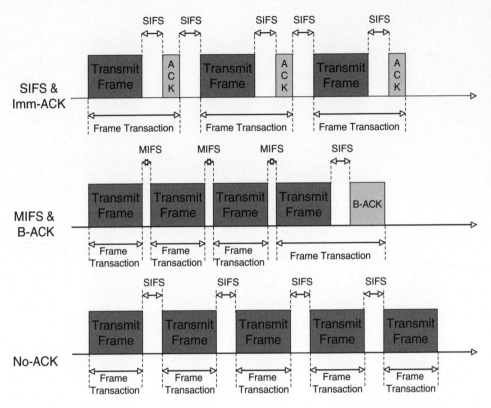

Figure 4.23   Examples of frame transactions with SIFS, MIFS, Imm-ACK, B-ACK, and No-ACK

At the receiver, the MAC sublayer is supposed to reassemble the fragments correctly back into the original MSDU before delivery. If a fragment is not received properly, then, depending on the ACK policy being used, the receiver may either discard the MSDU in entirety or wait for retransmissions. If retransmissions are not received in a timely fashion, then the receiver may discard the MSDU as well. Here, 'timely' is not defined in the specification and, therefore, is left to implementers to choose.

Conversely, up to 63 MSDUs may be aggregated by the MAC sublayer into a single MPDU, as long as their Delivery IDs are identical. The maximum size of an aggregated payload is still 4095 bytes. See Section 4.1.13 and Figure 4.22 for details on the format of the Aggregated Data frame. Note that the aggregated MSDUs must be aligned to 4-byte word boundaries. This will help with implementations on 32-bit buses.

In any case, the fragmentation/aggregation has to be transparent to the MAC client – as if no fragmentation/aggregation took place.

4.5.5 Channel Selection

Channel selection refers to the task of choosing a logical channel – a band group plus a TFC – in which the device is to operate. At first glance, this should be an implementation issue and not for standards to decide. After all, a device should be able to operate in any band group and TFC it chooses so long as it is following the rules and regulations associated with that choice. And, in fact, this *is* for the most part left to implementation in the MAC specification.

However, if the big picture is considered, then some level of agreement among devices may be necessary so that they can find each other. After all, what is the point in buying a UWB device if it has trouble finding like devices with which to communicate. With six band groups defined so far, and 10 TFCs in most of them, there are over 50 different logical channels to search through to find other devices. And, if those devices are also scanning all the logical channels, then the probability that they will find each other in a timely fashion will be quite small.

Consequently, the need for a channel selection process by which devices can detect each other's presence seems very much necessary. Unfortunately, this process is not in the current MAC specification. One may argue that this is not in the jurisdiction of the MAC specification anyway. Nevertheless, currently, devices are on their own to come up with schemes to find each other. There will, no doubt, have to be a revision in the near future coming out of WiMedia to address this critical issue.

Another aspect of channel selection is related to what is called Dynamic Channel Selection in the WiMedia MAC. This topic is related to the procedures a group of devices must follow if they would like to change channels in concert. The process starts with one device deciding to change channel (band group–TFC combination). It must, of course, first scan the channel of interest as dictated by the normal Beacon Group joining procedures (see Section 4.8). Then it will announce its intention to change channel by including in its Beacon in the current channel a Channel Change IE (see Table 4.3), in which it will include the channel number of the new channel as well as the Change Channel Count value indicating the number of superframes before the change will take place. All neighbors hearing such a Channel Change IE and desiring to move with this device to the new channel may then add the same IE with all the same field values in their Beacons. After the countdown reaches zero, all such devices will move to the new channel in the same superframe.

There are some ambiguities left in the specification regarding Dynamic Channel Selection, however. For example, the coordination of the relocation of the devices to the new channel should be improved such that they do not end up colliding with each other while trying to join the new channel. If all devices move to the new channel in the same superframe, then there would conceivably be a high chance of collisions among their Beacons, especially if the number of moving devices is larger than the number of unoccupied Beacon Slots in the new BP. Of course, the MAC protocol allows for recovery from such collisions, so the stability of the network is maintained.

Thus, this section of the specification may require more work in the next revision. At the same time, it can be argued that channel selection/relocation can be coordinated at the application level and need not be meticulously defined or controlled in the MAC sublayer.

4.6 Synchronization

4.6.1 Sources of Timing Error

As should be obvious by now, the WiMedia MAC is highly dependent on a tight control of timing synchronization among devices in the network. The (as defined in Section 4.2 and illustrated in Figure 4.5), specifically, is the point of synchronization for all peer MAC sublayers. Superframe and MAS timing will be useless if devices do not have a way of keeping a synchronous BPST time-frame among them.

On the other hand, clock drift is an unavoidable phenomenon in electronic circuitry. Moreover, the slower the drift in the clock requirement is, the more expensive the crystal has to be, making the device more expensive to build. Thus, in order to make a compromise (timing accuracy versus price), the WiMedia Alliance set a basic specification of ± 20 ppm on the clock accuracy of all WiMedia devices. (Any two devices could have up to 40 ppm difference in their clock frequencies.) This means that for every 1 s of time passing by, a maximum of 40 μs time drift is allowed. Hence, for a superframe, 65.536 ms, this drift can be calculated to be

$$\text{Max. allowed drift per superframe} = 40 \times 10^{-6} \times 0.065536 = 2.62 \text{ μs}$$

Unfortunately, the drift does not end with the end of a superframe. It accumulates indefinitely.

In addition to the clock drift, there are other sources of timing error that need to be considered. One is the internal device timing resolution. At the MAC sublayer, the clock accuracy resolution is 1 μs. This means that any timing information kept in the MAC could have up to ± 1 μs of error.

The other source of timing error that could be considered is the propagation delay. Fortunately, the short range of the UWB signals makes this source negligible (on the order of nanoseconds) compared with the above two.

In summary, the timing resolution and the timing drift will be the main sources of synchronization error among devices. In order to combat them, the WiMedia MAC provides two mechanisms: Guard Time and BPST Alignment.

4.6.2 BPST Alignment

As mentioned above, there could be a maximum of about 2.62 μs inter-device clock drift per superframe. That is, in the worst-case scenario, if left unchecked, the BPST of two devices in a neighborhood could drift away from each other at the rate of 2.62 μs per superframe. The Guard Time, as discussed in Section 4.6.3, enables the device to handle this drift up to a certain point. However, as the accumulation grows, the devices will eventually be unable to maintain their timing synchronization and the network will fall apart.

To prevent the accumulation of the clock drift, the MAC sublayer requires that every device resynchronize its internal timers every superframe. The reference time for the superframe is the BPST. All other counters and timing information is supposed to be derived from the BPST timer. Therefore, every superframe, all devices must synchronize their BPST.

To enable such synchronization, the MAC requires every device to maintain a table of timing differences between the actual arrival time of each neighbor's Beacon and the expected arrival time of that neighbor's Beacon. The expected arrival time is calculated based on the Beacon Slot number and its distance in time from BPST. This difference table may contain positive or negative values for each neighbor. If it is positive, then the neighbor's clock is slower than that of the device, and vice versa if negative.

Each device maintains its synchronization with its neighbors by simply synchronizing to the slowest neighbor in the neighborhood. The device does this by delaying its BPST by the largest positive value in its timing difference table. However, if the delay is larger than 4 μs, then the delaying process has to be spread over multiple superframes such that the adjustment in any given superframe is no more than 4 μs. This latter limitation prevents the loss of synchronization with other neighbors while trying to synchronize to the slowest neighbor.

The action time of this adjustment within the superframe is left to implementation, but it has to take place before the beginning of the next superframe. The synchronization should be applied even if the Beacon from the slowest neighbor is lost from time to time, and for up to three superframes in a row. A device should assume the slowest neighbor with missing Beacon is still around and use historical data to synchronize its BPST to that neighbor. After three superframes of lost Beacons from a slowest device, the next slowest neighbor is to be used for synchronization.

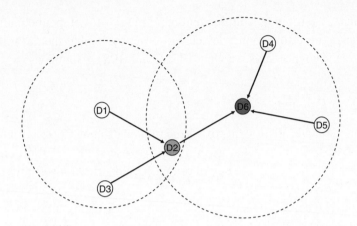

Figure 4.24   Example of BPST alignment in a multi-hop network

In a multi-hop network, the BPST alignment may get more complicated, but the methodology described above still works. For example, consider the two-hop network of Figure 4.24, which shows the one-hop neighborhoods (Beacon Groups) of devices D1 and D6 in dashed circles. (Note that the concept of neighborhood or Beacon Group is always relative to a device, since each device may see a different set of neighbors.)

Let us assume that in the Beacon Group of D1, D2 is the slowest device (among D1, D2, and D3), and in the Beacon Group of D6, D6 is the slowest (among D2, D4, D5, and D6). The arrows in Figure 4.24 indicate to which device each neighbor synchronizes its BPST. As seen in the figure, devices D1 and D3 synchronize to D2, while D2 itself, along with D4 and D5, synchronize to D6. Thus, in any given superframe, as the relative clock drifts of D6 and D2 are measured by the other devices, D1 and D3 change their BPSTs to match that of D2 but, at the same time, D2 changes its BPST to that of D6. As time goes by, this process continues indefinitely (barring any change to the neighborhoods). Hence, D1 and D3 never quite align their BPST to D2's, but the difference is always kept within the tolerable limit (4 µs). Consequently, this cat-and-mouse chase may continue indefinitely but without disruption to the network operation.

Once a device goes into hibernation, per-superframe synchronization is no longer possible. Therefore, it is quite reasonable to expect that, when the device wakes up from hibernation, it is completely out of alignment with the rest of the network. Consequently, the MAC specification requires every hibernating device to wake up one superframe earlier than when it has to send a Beacon and to use that superframe to scan the Beacons and subsequently synchronize itself to the slowest neighbor.

4.6.3 Guard Time

So far, we have seen that the MAC sublayer is capable of maintaining inter-device superframe timing to within about 2.62 μs. However, this is true only if every device's Beacon is heard by all neighbors every superframe. Of course, this cannot be guaranteed in a wireless channel. For this very reason, the MAC protocol allows missing Beacons in a BP up to three superframes in a row. That is, if a device's Beacon is missing from a BP, the neighbors may not assume that the device has left the neighborhood until at least four superframes have passed. Consequently, a clock drift may accumulate up to four superframes before there is a chance to correct it. This means that the maximum clock drift to handle is actually

$$\text{Max. clock drift} = 2.62 \ \mu\text{s/superframe} \times 4 \ \text{superframes} = 10.49 \ \mu\text{s}$$

In addition to this drift, as mentioned in Section 4.6.1, there is a maximum of a 2 μs timing resolution error inherent to the MAC sublayer. Adding this to the Max Clock Drift, we get

$$\text{Max. timing error to handle} = 10.49 \ \mu\text{s} + 2 \ \mu\text{s} \approx 12 \ \mu\text{s}$$

The MAC sublayer, therefore, requires a 12 μs Guard Time in addition to the 10 μs SIFS duration to be allowed by each device at the end of each reservation block (including Beacon Slots). This way, in the worst-case scenario that the timing of a device is off by the maximum amount (12 μs), there will be sufficient time to allow one reservation block to end and at least an SIFS time (10 μs) to pass before the reservation block of another device starts. The inter-device timing error (up to 12 μs), therefore, is not removed by the Guard Time. Instead, all reservation blocks are spread far enough apart from each other so that no collision of packets may occur due to any potential synchronization errors between devices.

To exemplify the effect of Guard Time, consider the scenario depicted in Figure 4.25. Device 1 (D1) has a slower clock than Device 2 (D2). Therefore, there is a timing offset between the two devices. (D1's MAS boundaries do not match those of D2.) D1 has a reservation block that is followed by D2's reservation block. During its reservation block, D1 performs frame transactions with its own target device (not necessarily D2). Recall that frame transactions are described in Section 4.5.3. In between frame transactions, all D1 has to allow for is the SIFS timing. However, at the end of the reservation block, D1 has to make sure that there is at least an SIFS timing between D1 and D2. Since D1 is not sure of its timing offset with D2, it has to provide for a Guard Time as well. If D1 does not allow for Guard Time, then D2 could initiate its frame transaction partly into the SIFS timing of D1. Since it is important to leave at least an SIFS timing between any two

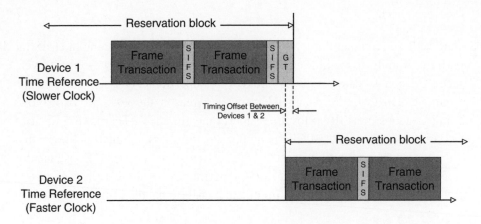

Figure 4.25   Guard Time illustration

frame transactions, the inclusion of the Guard Time at the end of the reservation block is necessary.

Note that Guard Time is not necessary in between frame transactions within a reservation block. This is because, in this case, (a) the timing kept is only between D1 and its target(s) and (b) critical timing is relative to transmitted or received packets (not absolute time). That is, each target knows when the end of a received packet from D1 is and, thus, can count up to SIFS time before sending any necessary acknowledgments followed by another SIFS time, before the next packet starts arriving. During a SIFS time, any time drifts are negligible.

By the same token, since Beacon packets are kept apart based on the absolute time of the Beacon Slot boundaries, the inclusion of a Guard Time (in addition to SIFS) within a Beacon Slot is necessary. Guard Time plus SIFS adds up to 10 μs + 12 μs = 22 μs, taking precious payload space out of Beacon Slots, as calculated in Section 4.4.1. Nonetheless, the Guard Time is a very cost-effective method of avoiding collisions among devices due to clock drifts, as long as BPSTs are aligned and there is no more than a 20 ppm clock drift in each device.

4.7 Power Conservation

Power conservation is a very important aspect of UWB, in general, and specifically in WiMedia. The ability of UWB to deliver high-throughput data transfers at much lower power consumption than competing technologies is the driving force behind WiMedia specifications. Of course, implementation choices make a great deal of difference when it comes to power saving. However, if a protocol is not power sensitive, then, no matter how power conscious the implementation is, the overall power consumption will be high.

WiMedia MAC places a special emphasis on power saving. There are two levels of power savings provided in the specification. The first differentiates between an *Active* mode and a *Hibernating* mode. The second defines, for the Active mode, two different states of *Awake* and *Sleep*.

Implementations may add finer granularity to their use of these modes and states in reducing device power consumption. For example, they may define different levels of power consumption depending on what component of the device (RF, Baseband, MAC Hardware, MAC firmware, Input/Output interface, etc.) is turned off or on in certain conditions. The choices are numerous and outside the scope of this book.

Let us now examine each of the MAC-defined power modes/states.

4.7.1 Power Modes

Two power modes are defined in WiMedia MAC: Active and Hibernation

The Active mode is defined by the MAC specification as one in which the device will send and receive Beacons in the current superframe. In other words, if a device is actively participating in the BP (sending its Beacon and receiving all other Beacons) as it should, then the device is considered to be in the Active mode for the whole superframe.

As opposed to the Active mode, the Hibernation mode is one which allows the device to retire from the transmission of frames of any sort (Beacon or otherwise) in one or more superframes. In fact, the official definition of a device being in Hibernation mode in a superframe is that the device does not send any frames in that superframe.

This is the mode in which the device may be able to save the most power by shutting down most of its circuitry, even possibly its main clock. However, in order to prevent problems arising out of transitioning from Active to Hibernation or from Hibernation to Active modes without proper preparations, there are certain rules that need to be followed.

Going to hibernation is relatively simple. A device announces its intent to do so by including a Hibernation Mode IE (see Table 4.3) in its Beacon several superframes ahead of actual hibernation. The IE determines the duration of hibernation and a countdown value to the hibernation. Every superframe, the countdown value is decreased by one until it reaches zero. Then, the hibernation starts in the following superframe.

If there are any DRP reservations left open by the hibernating device, then they will be assumed canceled when the device goes into Hibernation mode.[14] During

---

[14] As a 'good neighbor,' a hibernating device should have closed out its DRP reservations before hibernating.

the hibernation period, all active neighbors of the hibernating device will mark its Beacon Slot as occupied and nonmovable in their BPOIE. All traffic destined to a hibernating device must be buffered until the device comes back to the Active mode, unless the traffic is a multicast one, in which case it may be transmitted and/or buffered.

Once in hibernation, a device must make no transmissions of any sort. Although it announced its hibernation duration, the device may return from hibernation at any time, as long as it scans for one or more superframes ahead of going into Active mode in order to:

- Resynchronize its BPST as required by Section 4.6.2.
- Determine what Beacon Slot to use for transmitting its Beacon:
  – if the previous Beacon Slot is still available, the device may use it;
  – Otherwise, the device has to use the normal rules of joining the Beacon Group.

One optional feature of the WiMedia MAC is the provision for a Hibernation Anchor, defined as an Active device that accepts the task of announcing to the neighborhood the pertinent information about the hibernating device, while that device is in hibernation. This information is included in the Hibernation Anchor IE of the anchor device. Devices that join the group during the hibernation can use this information to learn about the existence of the hibernating device, and its hibernation schedule, so that they can prepare for communication with the device when it returns to the Active mode. Once the hibernating device is back in the Active mode, the anchor stops sending the Hibernation Anchor IE.

4.7.2 Power States

Once in Active mode, a device is considered to be in one of two states: Awake or Sleep.

The Awake state of the WiMedia MAC is not quite what it intuitively means. If Awake, then the device is ready to transmit or receive. Thus, while in the Awake state, the device has all its components (especially PHY) powered on and ready. If the device is not transmitting in this state, then it is attempting to receive.[15]

A device may go to Sleep if it does not have anything to transmit or receive. However, the device must make sure it goes into the Awake state at least a Guard Time before it actually needs to receive a packet. This is because of the potential clock drift that may exist between devices, as described in Section 4.6.3. Thus,

---

[15] Attempting to receive includes the act of searching for incoming packets (acquisition) even if there are none coming.

12 μs before BPST or a scheduled DRP reception, the device must be in the Awake mode (i.e. attempting to receive the packets 12 μs ahead of schedule just in case the transmitting device has drifted up to 12 μs early).

Consequently, all devices must go to the Awake state 12 μs prior to BPST and stay there until the end of the BP. Also, all devices that have DRP reservations must be Awake 12 μs prior to the start of the DRP period and stay Awake during the period of the reservation. An exception to this rule is when a transmitting device has no more data to transmit during its DRP reservation. Then, it may go to the Sleep state. Optionally, releasing the rest of the DRP reservation (using the Unused DRP Reservation Announcement frame) would indicate to the receiver of the reservation that the receiver can go to Sleep as well.

Note that if the device is expecting a PCA reception (see Section 4.10) it has to be Awake 12 μs prior and all the way through the MASs it announced in its PCA Availability IE (see Table 4.3). If no PCA Availability IE was sent in its Beacon, then the device must stay Awake through all of the PCA MASs in the superframe. Therefore, it behooves the device to announce its availability for PCA communication in its Beacon so that it does not have to waste power unnecessarily.

4.8 Beacon Protocol

Let us delve into one of the most interesting parts of this MAC, the Beacon protocol, which, in the context of WiMedia MAC sublayer, refers to the rules associated with how the Beacons are generated and used by each device in order to maintain a distributed control of the physical medium among the devices. Since the medium is a wireless channel and the devices may be mobile, there are certain challenges in this control mechanism; hence, the Beacon protocol has to be intelligent enough to address the associated stability and efficiency problems that may arise dynamically in different circumstances. Usually, intelligence translates to complexity in design; hence, the Beacon protocol is one of the more complicated parts of the MAC sublayer. However, through the use of this protocol, the WiMedia MAC promises to be one of the most efficient MAC protocols and provide a true ad hoc, peer-to-peer network offering guaranteed QoS.

4.8.1 BP

Through the broadcast of control information during the BP, the devices are able to discover each other and each other's capabilities, manage their data communication, monitor each other's fair access to the channel resources, allow for mobility within the network, etc. The BP is defined loosely as the period at the beginning of each superframe in which Beacons are expected to be transmitted. This period is not fixed; it changes with the number of devices in the extended neighborhood. In an ad hoc, peer-to-peer network, defining such a dynamically changing BP is not an easy task. Consequently, there are specific rules that each device needs to follow in calculating the BP length. Each device must also make sure it is coordinating its calculations with the other devices' BP lengths.

There is an absolute maximum length to the BP, however: 96 Beacon Slots, equivalent to almost 32 MASs. That is, in the MAS table of Figure 4.6, the first two zones (columns) can potentially be used for Beacon Slots. However, practically, this is a very rare event to be expected. For one thing, under normal conditions (without BP merging), the maximum a BP length can grow to is 48 (one MAS zone). Only when two neighborhoods[16] of 48 Beacons merge together (to be explained later) can the BP length grow to its maximum size. Even then, it is hard to imagine, at least for now, that 96 devices or applications are able (or even need) to coexist in a single neighborhood. Additionally, once you take away two zones out of the 16 zones of Figure 4.6, there does not seem to be much capacity left in the medium (on a single channel) for 96 devices to communicate effectively with each other, forcing some of them to move to a different channel anyway. Hence, it is very unlikely to end up in a scenario with more than 48 Beacon Slots (one zone) to be used. Nevertheless, the maximum of 96 is there to take care of the extreme corner cases, should they arise.

As described in Section 4.2, the beginning of every superframe is dedicated to the transmission of Beacon frames, which coincides with the beginning of the BP. Recall that each Beacon Slot is exactly 85 μs in length and all Beacon packets along with their Guard Time and SIFS must fit within that slot, making the maximum size of the Beacon payload 320 bytes (see Section 4.4.1). Figure 4.5 showed us how Beacon Slots are superimposed on top of the MASs; and depending on the number of devices in the neighborhood, the Beacon period could take up more or fewer MASs.

Naturally, as long as a BP is defined to overlap a certain MAS, then that MAS or part thereof may not be used for any purpose other than the broadcast of Beacon frames. There are scenarios in which a BP could expand into a device's reservation block. Even in those cases, the Beacon transmissions take priority over any other frame types. That is, as long as the BP is covering a DRP reservation, the reservation for the overlapping MASs (or partial MASs) are considered void and the associated devices must avoid using those MASs (or partial MASs) for any purpose other than Beacon transmission/receptions.

---

[16] A neighborhood is defined as all devices within range of each other (one-hop or multi-hop range) that have the same BPST.

The Beacon Slots in a BP are numbered sequentially, starting from zero. The first two Beacon Slots (slots 0 and 1) of any device's BP[17] are considered special. Called Signaling Slots, they are a set of reserved slots for when a regular Beacon Slot is not available for a device to use in any of its neighbors' BPs. The device would then transmit a Beacon frame in one of the Signaling Slots (randomly chosen between the two slots to reduce the chance of collision with Beacons from other devices that need to use the Signaling Slots). This indicates to the other devices that the BP needs to be extended to accommodate the new device. Hence, a device does not normally transmit its Beacons in Beacon Slots 0 or 1, i.e. the Signaling Slots. Slot 2 is the first regular Beacon Slot of any BP.

Before any initial transmission of any type of frame, a device is expected to scan the channel of interest for potential Beacons from other devices. This scan must be at least one superframe long. If no Beacons are discovered, then the device may consider the channel empty and start its own BP. This means that the device may define its own BPST and send its Beacon in Slot 2 (the first slot after Signaling Slots). This case produces the shortest BP length: three (if the device adds no empty slots to the BP after its own slot), although typically devices add two to three more empty slots to let others join in. Thus, a typical BP for a device that has no other neighbors would be two MASs (six Beacon Slots) long.

On the other hand, if the scan of the channel uncovers valid Beacons[18] from one or more devices, then the device may not start a new BP. Instead, it must join the neighborhood.[19] Once joined in, an Active[20] device is expected to listen for all Beacons transmitted during the BP it announced in the last superframe, including the ones in Signaling Slots. In certain cases, the device is expected to increase its BP length (but never by more than eight slots, and never beyond the maximum BP length). For example, when a device receives a Beacon frame during the Signaling Slot, it must extend its BP length to include the Beacon Slot indicated by that frame.

Each device is expected to announce in its Beacon frame its own vision of what the BP length was in the previous superframe. The length must cover all the occupied Beacon Slots of the previous four superframes, as seen by the device from all its neighbors, plus, optionally, a small number of extra slots at the end for any new device to join into the neighborhood. The extra slots may not be more than eight, but may be none. Since devices may come and go, a BP may end up with Beacon

---

[17] Note that each device may have a different impression of what the BP length is.

[18] Note that if a device receives an invalid Beacon frame (a frame with an invalid FCS), then it should scan for one more superframe before taking any action.

[19] A neighborhood is defined as all devices within range of each other (one-hop or multi-hop range) that have the same BPST.

[20] One that is not hibernating.

Slots that are sparsely occupied by devices. To prevent a perpetual expansion of the BP, BP Contraction mechanisms (see Section 4.8.3) are provided in the WiMedia MAC.

For a new device to join a neighborhood, it must transmit its first Beacon in one of the eight Beacon Slots after the last unavailable Beacon Slot. If this puts the slot beyond the Beacon Length announced by any of the neighbors, then the device must also send its Beacon frame in one of the two Signaling Slots. The only difference between the Beacon frame sent in the Beacon Slot and the one in the Signaling Slot is that the latter has its Signaling Slot bit set to 1 (see Figure 4.9). The device will continue using the Signaling Slot until all the neighbors extend their BP length to include its Beacon Slot. However, in this situation, the use of Signaling Slot may not be continuous. Every four superframes, the device has to stop Signaling Slot transmissions for four superframes before resuming. This allows for other devices to have a chance to use the Signaling Slot if needed.

A Beacon Slot is available for occupancy by a device only if in the last four superframes the Beacon Slot was not reported as occupied by any of the transmitted or received BPOIEs. For a description of BPOIEs, refer to Section 4.4.1.1. Note that by requiring the device to check the BPOIE of all its neighbors, in effect, the device is checking the Beacon Slot occupancy for a two-hop neighborhood (the neighbors' neighbors). This is important because, without checking the extended (multi-hop) neighborhood, the MAC runs the risk of having the hidden-node problem, as described in Section 4.1.4.

4.8.2 Beacon Loss or Collision

From the discussions above, it is obvious that Beacons are extremely important to the operation and stability of the MAC protocol. However, since the devices are expected to operate in unpredictable and nonstationary environments, it is quite reasonable to assume that, from time to time or in certain circumstances, Beacons could be lost or they could collide. Beacon loss means the nonreception of an expected Beacon frame during a Beacon Slot. On the other hand, Beacon collision refers to the transmission of Beacon frames by two or more devices in such timing that they fully or partially overlap each other, causing one or more of the Beacon frames not to be received properly. From a MAC perspective, Beacon collisions are very undesirable but unavoidable. Unfortunately, detecting a Beacon collision is not simple, mostly because the symptom is the same as that of severe channel impairment: loss of packet. Thus, the MAC has to be most vigilant on detecting Beacon collisions. Once detected, the remedies are usually straightforward.

When it comes to Beacon loss, the MAC protocol allows for missing Beacons in three consecutive superframes from a previously transmitting neighbor, after which time the corresponding Beacon Slot in considered unoccupied. Conversely, if a device is heard during the BP at least once every four superframes, then the neighbor is still alive and its Beacon Slot intact. This is done to prevent instability in the Beacon protocol due to fading conditions in the wireless channels. When a device's Beacon is lost but for not more than three superframes, the immediate neighbors (one-hop neighbors) are expected to use the most up-to-date information on the missing device. This means that each device needs to maintain a database on their neighbors for at least three superframes and to update it whenever they get a new Beacon from them.

In WiMedia MAC, the Beacon collision detection is achieved by any and all means available to MAC. Here is a list of them.

1. By requiring that every device stop transmitting its Beacon once in a while and instead listen for transmissions within its own Beacon Slot. This is called Beacon skipping and is done nonperiodically in order to avoid the corner case of the colliding devices choosing the same skipping period. The skipping takes place at least once every 128 superframes. If the device detects even a hint of a transmission during these Beacon skips it will assume there has been a collision. These hints include:
   (i) PHY indicating there was medium activity but no error-free MAC header;
   (ii) MAC sublayer receiving a Beacon frame header.
2. By listening to what the neighbors say about its Beacon Slot. In the BPOIE of each neighbor, there is information about the device and its BP occupancy. For example, if any of the following is noticed, then a Beacon collision is declared:
   (i) the DevAddr corresponding to the device's Beacon Slot is not that of the device, nor is it BcstAddr (broadcast address);
   (ii) after a Beacon skip, the BPOIE of the neighbors indicates that the skipped Beacon Slot was actually occupied.

4.8.3 Beacon Contraction

Initially, as the BP is getting populated by new devices joining the neighborhood, the Beacon Slots are occupied more or less sequentially. The BP at this point may look well packed, with few unused Beacon Slots. However, as time passes and devices leave the neighborhood and others join, and as neighbors' neighbors influence the BP's occupancy, it is reasonable to imagine that the BP could grow large but become sparsely occupied. This is not desirable for two reasons. First, the longer the BP becomes, the fewer usable remaining MASs in the superframe will remain for actual data communication, reducing the overall throughput of the channel. Second, the power-sensitive devices would have to spend longer periods of time listening (keeping their PHYs powered on) for Beacons in slots that are otherwise empty. Any unnecessary reception activity would use up precious battery power in such devices.

Thus, it is a good idea to make the BP as tightly packed as possible. However, this is not possible without some MAC-level maintenance activity. This activity is called BP Contraction. The contraction must be done in such a way so as not to disturb the balance of stability (in fading channel) versus efficiency (throughput and power). That is, the contraction should not happen too fast or too slow. Plus, since there is no master in this MAC protocol, the distributed control of the MAC sublayer has to make sure all devices follow a collective pattern of contraction in an orderly fashion.

In order to accommodate all of these requirements, the WiMedia MAC has the following contraction procedure. First of all, each device must determine if its Beacon is movable. A device marks its Beacon movable if it can find an available Beacon Slot earlier in the BP than its own (with a Beacon Slot number smaller than its own). Note that, as shown in Figure 4.9, there is a Movable Beacon bit in the Beacon frame payload specifically for this purpose. There are, of course, exceptions to this rule. For example, if the device is planning on going into hibernation (if it is including Hibernation Mode IE in its Beacon frame payload), then the Beacon is not movable. Also, a device must not be involved in Beacon collision (see Section 4.8.2) or BP merge (see Section 4.8.5) to follow BP contraction.

To make the moves orderly (and to prevent unnecessary Beacon collisions due to simultaneous moves to the same Beacon Slot), the contraction must take place one device at a time. Also, to prevent hysteresis, a device must have a movable Beacon for at least four superframes before it is expected to take part in contraction. Then, the device with such a movable Beacon in the highest Beacon Slot number has to move first. It relocates its Beacon to the earliest unoccupied Beacon Slot. Once this happens, the next device that finds its Beacon closest to the end of the BP will find itself in the position to potentially move its Beacon, and so on. Consequently, slowly but surely the BP contracts, as long as no other events (e.g. new-coming devices, BP merge, etc.) happen to expand the BP again.

4.8.4 Alien Beacons

As mentioned previously, the WiMedia MAC has to cope with quite a number of challenges, including the wireless channel and the mobility of the devices. As such, it needs to be well prepared for unplanned (and undesirable) conditions. One such condition, mostly caused by mobility, is the possibility of having two distinct neighborhoods of devices suddenly finding themselves within range of each other.

Such a condition would result in having two BPs and two different BPSTs (i.e. two different superframe timings) to be used by the two groups. Once these two groups of devices can hear each other, all the MAC structure of superframes, MAC slots, reservations, etc., would potentially be useless. Collisions could start to take place among Beacons or other frames. If nothing is done about it, the MAC protocol would start to fall apart.

To avoid such a doomsday scenario, the WiMedia MAC is very diligent in finding and advertising what it calls Alien Beacons. A device declares a validly received Beacon frame (with correct HCS and FCS) from a neighbor as an Alien Beacon if it indicates a BPST that is different from that of the device's own BPST.

Of course, there are exceptions to this rule as well. For example, a small BPST offset should not cause an Alien Beacon declaration. Clock drifts are normal among devices, and some drift faster than others. Thus, as long as the BPSTs of the neighbors are within twice the Guard Time of each other, they are considered as normal. Of course, the BPST alignment procedure (see Section 4.6) will make sure these normal BPST offsets do not grow to become abnormal. Another exception is when a Beacon with a Signaling Slot bit set to 1 is received.

Once such an Alien neighbor is identified, the whole BP defined by that neighbor's Beacon is considered an Alien BP. Once devices notice an Alien BP, they are expected to take remedial actions. These involve BP merging, among other things, which will be discussed next.

4.8.5 BP Merging

Since Alien Beacons can have a devastating effect on the performance and stability of the network, the MAC protocol requires devices in different neighborhoods that come into RF range of each other to merge their BPs. In an ad hoc, peer-to-peer network, taking on the challenge of merging two uncoordinated neighborhoods of devices is not a 'walk in the park.' The challenge is to get all devices of both groups to follow the same procedure without knowing if all of them are listening to each other. As a result, BP merging is one of the more complex procedures in the WiMedia MAC.

When one or more Alien BPs are detected, their relative timings could be such that they either overlap the existing BP or not. In either case, the BPs need to be merged (their BPSTs aligned). However, there are different rules for addressing overlapping and nonoverlapping BPs. Also, once the BPSTs are aligned, all the Beacons of the merging BP must be accommodated in the merged-to BP as orderly as possible. Plus, there are certain special cases that need to be handled properly, such as when the devices in a neighborhood have no clue that there is an Alien BP to which they need to move.

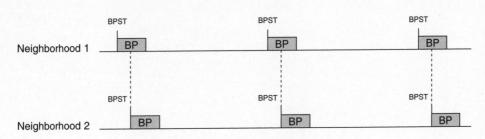

Figure 4.26   Example of an overlapping Alien BP

To make sure the merge is performed in an orderly fashion among the neighborhoods, the devices must decide whether they are to merge their BPs with the other (Alien) BPs or wait for the Alien BPs to merge with theirs. Note that devices belonging to each BP consider the other BPs as Aliens. The choice of which neighborhood moves and which stays is based on the relative timings of the two BPSTs. If two BPs overlap in time, then the device with a BPST that falls within the Alien BP shall move. See Figure 4.26 for an example of an overlapping Alien BP. In this example, all devices belonging to Neighborhood 2 are expected to move their BPSTs and merge their Beacons with those of Neighborhood 1.

If the BPs do not overlap, then the determination is done based on whether or not the BPST of the device falls in the first half of the Alien superframe. Figure 4.27 illustrates an example of such a scenario. Typically, the devices with BPSTs in the first half of the Alien superframe move to the Alien BP. However, there are a few more twists to this rule that will be explained later in this section.

There is a major difference in the move with overlapping BPs relative to one with nonoverlapping BPs. The difference is that the overlapping BP is considered more detrimental to the health of the MAC protocol and requires an immediate action. When two BPs overlap, it is very likely that some or all of the Beacons in both BPs are colliding, resulting in no MAC management of the medium. This is very dangerous and should be remedied immediately. That is why, with an overlapping Alien BP, the move takes place immediately. During the relocation, the moving

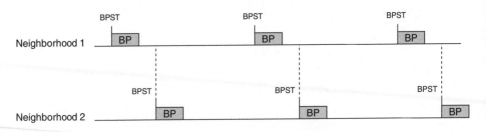

Figure 4.27   Example of a non-overlapping Alien BP

devices have two choices for a new Beacon Slot number. The first one is to follow the normal procedure for joining a neighborhood. The other is to keep the relative Beacon Slot numbers of the joining BP by simply appending all the Beacons to the end of the join-to BP.

On the other hand, when the Alien BP is nonoverlapping, the likelihood of Beacons colliding is very low (they could still collide with normal traffic frames, but not with other Beacons). Consequently, the move can take place in a more orderly fashion and with more planning. In such a case, the device is expected to announce the discovery of the Alien BP. The announcement is made by making an Alien BP reservation (a DRP IE with Reservation Type set to Alien BP) over the MASs that completely cover the Alien BP. This announcement has two benefits. For one, it disallows any other device from making use of those MASs for normal traffic and, hence, it protects the Alien Beacons from colliding with normal traffic. Second, it tells all other devices in the Beacon Group about the Alien Beacon so they all can react to it as needed.

Next, the device must decide if it has to move to the Alien BP or to wait for the Alien Beacons to move to its BP. The MAC rule is that, within the first 128 superframes, the device with the Alien BPST falling in the first half of its superframe shall relocate its Beacon to the Alien BP. Once it is determined that it shall relocate its Beacon, the device includes a BP Switch IE (see Table 4.3) in its Beacon to announce its intention to relocate its Beacon to the Alien BP. In the BP Switch IE, the device includes a counter value of 9 that counts down every superframe the time to actual switch. The BP Switch IE also contains a BPST Offset field that announces the largest difference between the current BPST and any of the Alien BPSTs. These announcements further serve the purpose of letting all other devices know that a move is taking place, and the countdown helps with planning the smooth transfer of the traffic flow. This is especially helpful if there is more than one Alien BP involved.

By noticing the BP Switch IE, all moving devices of all BPs can coordinate their moves to the final resting BP without requiring multiple intermediate moves. Any device that hears the BP Switch IE must also include one in its Beacon. This will make sure that all devices in the extended neighborhood will hear about the move to the new Alien BP. The countdown value for all devices will be copied identically so that all devices move at the same superframe.

The Beacon Slot to be chosen by each moving device can again be based on a concatenation of the current BP with the Alien BP, or, alternatively, the device can simply use the normal join rules to join the Alien BP. The device indicates this choice in its BP Switch IE in the Beacon Slot Offset field (nonzero for the former and zero for the latter choice). At the join-to BP, if the devices hear an Alien BP (from the join-from BP) with BP Switch IE and a Beacon Slot Offset field set to a nonzero value, then the devices are to extend their BP lengths to at least the sum of the two BPs (but always less than or equal to the maximum of 96 Beacon Slots). This makes sure that the BP is extended sufficiently to receive the new Beacons from the Alien BP.

While the countdown is taking place, if new information about the Alien BP or the relocating neighbors surfaces, then the device may have to adjust or even halt and restart its relocation process. There are quite a few corner cases that are meticulously covered in the standard to make sure all conditions are handled properly to allow for an orderly and efficient relocation.

As mentioned above, if a device determines that it has to relocate its Beacon, it must do so within 128 superframes. However, there is always a chance that the device never realized that there was an Alien BP in the first place. For example, let us consider the condition where a device detects an Alien Beacon and does all it can to advertise and protect it in its own BP, but, based on the relocation rules, it determines that it is not to move. On the other hand, no devices in the other BP have discovered their Alien BP and no relocation is taking place. To cover such a case, the MAC rules indicate that, after waiting for a relocation of the Alien BP to the current BP, the device that discovered the Alien BP can initiate a move to the Alien BP instead. The wait time in this case is 192 superframes.

4.9 DRP

As introduced in Section 4.1.5, a DRP is a reservation protocol that uses a TDMA scheme of sharing the resources (the wireless medium) among the devices. This is in contrast to the CSMA with Collision Avoidance (CSMA/CA), which is the basis for the PCA option in WiMedia MAC, described in Section 4.10.

DRP allows devices to reserve a set of MASs for communication with other devices. They do this via DRP IEs they send in their Beacon frames (see Figure 4.10) or Command frames (see Section 4.4.3). Both the reservation owner (the device making the reservation and intending to transmit) and the reservation target (the device that is to receive the transmission from the reservation owner) have to send DRP IEs in their Beacons. The format of DRP IE is described in Section 4.4.1.2.

There are five reservation types that can be made with DRP IE. These are tabulated and described in Table 4.7. Note that, although DRP reservations were originally designed for noncontention-based communication, they later evolved to relate to PCA-type transactions as well. The relation is through the use of the DRP IE and the act of reservation of MASs. As seen in Table 4.7, even PCA communication may use reservations. It may sound strange to think that DRP IEs are used to reserve MASs for what will end up being a contention-based channel access.

Table 4.7 DRP reservation types

Reservation Type	Description
Hard DRP	Reserves MASs for exclusive use by reservation owner and target, and for a specific data stream (Stream Index). Any unused time in the reservation may explicitly be released (using UDA and UDR commands) for general PCA transactions. WiMedia MAC protocol must be strictly adhered to
Private DRP	Reserves MASs for exclusive use by reservation owner and target, unless unused time is explicitly released. The difference is that any application-specific channel access and frame exchange may be used in Private DRP, whereas WiMedia MAC protocol must strictly be used in Hard DRP. Frame formats and types must still follow WiMedia MAC. Private DRP is used, for example, to enable CW-USB protocol, which has a somewhat different MAC protocol
Soft DRP	Reserves MASs for PCA transactions with preference towards the reservation owner, which may access the channel with highest priority and without any backoff. As long as the owner maintains SIFS timing between its frames, it can continue to transmit without backoff. Otherwise, the neighbors of the owner (but not their neighbors) may get in the act by using normal PCA rules
PCA	Reserves MASs for PCA transactions. See Section 4.10
Alien Beacon	Reserves MASs to protect Alien BP. No device in the current BP may use the reservation except to send a Beacon in the Alien BP. See Section 4.8.4

However, there is some logic to this madness, and that is the fact that all devices benefit from the knowledge of when other devices may be interested in communicating via PCA. That way, devices may be able to save power by turning off their receivers during other parts of the superframe where there is no expectation of activity.

All devices are expected to decode all DRP IEs in all Beacon frames they receive during the BP and follow the instructions offered in those IEs. All neighbors must respect all reservations so that contention-free communication can take place.

To make sure the quality of service required by the application is maintained, a device must carefully consider the correct size and frequency of reservation blocks in a superframe. Examples of reservations are given in Section 4.4.1.2.

A reservation is not in place until both parties (owner and target) agree on it. The owner initiates the request for a reservation, but the target may not accept the reservation. The reason could be anything, but it is usually based on a conflicting reservation that the target may have with another neighbor. Since the owner may not see the target's neighbor, by negotiating with the target, the owner can make sure that the proposed reservation creates no potential interference to the target's neighbors, hence protecting against the hidden-node problem (as described in Section 4.1.4).

The DRP negotiation starts by the reservation owner sending a DRP IE with the following field values:

- Target/Owner DevAddr = DevAddr of the reservation *target*;
- Reservation Status = 0;
- Reason Code = Accepted;
- Stream Index = A new value (if a new stream), or an existing value (if an existing stream).

The target of a unicast or multicast reservation would then respond by setting its DRP IE parameters as indicated in Table 4.8. Note that the response to an owner's request for a reservation may be Granted, Not Granted, or Pending. If Not Granted, the target would give a reason of either Conflict or Denied. Conflict is an indication that there is a neighbor with a reservation overlapping the requested one. In this case, the target device must include a DRP Availability IE, which gives the MAS availability for that device. This will then help the owner to adjust its reservation request on its next attempt according to the target's MAS availability. A Denied response could be for any reason other than a conflict. If the response to the reservation request contains a Reason Code of Pending, it means that the target will need more time to respond with a final answer to the owner.

In all unicast reservation request responses, the target device sets the Reservation Status to 1 only if the request is granted by the target. Only then may the owner set its own DRP IE's Reservation Status to 1, which indicates to other devices that the reservation is now blocked for communication as requested by the owner.

Keep in mind that if the DRP negotiations are done implicitly (via DRP IEs sent in Beacons), there is only one Beacon per device in each superframe. Hence, every

Table 4.8 Parameter values for DRP IE of reservation target

If reservation is ...	Target/Owner DevAddr is set to ...	Reservation Status is set to ...	Reason Code is set to ...
Granted	DevAddr of the reservation *owner*	1 if unicast, same as owner's if multicast	Accepted
Not Granted	DevAddr of the reservation *owner*	0	Conflict, if there is a reservation conflict, Denied, otherwise
Pending	DevAddr of the reservation *owner*	0	Pending

response (from owner to target, or vice versa) will take at least one superframe (in the next BP) to show up over the air. Devices are expected to maintain their DRP reservation requests/responses (DRP IEs) in their Beacons for at least four superframes, or until they receive their responses. Once the unicast reservation is granted, the DRP IEs must continue to exist in the Beacons of the owner and the target, with Reservation Status set to 1 and the Reason Code set to Accepted, for as long as the reservation is necessary to keep.

Keep in mind that, owing to mobility and changing wireless channel conditions, and despite the best efforts of the MAC protocol to prevent it, devices may end up with conflicting (overlapping) reservations granted or being negotiated. By definition, PCA and Alien BP reservation types are not considered conflicting, even if they are overlapping. This is because PCA reservations are there only to announce the possibility of contention-based access by devices. Since PCA rules will be used during these reservations, contention (conflict) is expected and handled properly. As for the Alien BP reservation type, the point of the reservation is to protect the Alien BP, and since the MASs in those reservations do not normally get used, there will not be any conflict to worry about.

For the conflicting reservations it is very important for the MAC protocol to be able to resolve the conflict as soon as they are detected to reduce the possibility of frame errors due to collisions. Of course, as per the DRP negotiation rules mentioned above, if a device is a target of a reservation request that is overlapping with another reservation with the Reservation Status set to 1, then the device must set its own Reservation Status to 0 and set the Reason Code to Conflict, in the DRP IE.

If a device discovers in any superframe that its DRP IE and that of its neighbor have a reservation conflict (whether in negotiation or when granted), the conflict is resolved by considering the Conflict Tie-Breaker bit of the DRP IE as well as the Beacon Slot number of the device. The Conflict Tie-Breaker bit is set to a random value (0 or 1) by the reservation owner at the time of reservation request. All future negotiations and reference to that reservation by the owner or the target devices will maintain the same bit value in Conflict Tie-Breaker field. Once the reservation conflict is detected, the two devices will first compare each other's Conflict Tie-Breaker bit. If the two are the same, then the device with the smaller Beacon Slot number wins (keeps the reservation). If the two Conflict Tie-Breaker bits are different in value, then the device with the higher Beacon Slot number wins. In any case, the losing device must give up the reservation or its negotiation.

The losing reservation owner cannot request a new reservation right away. Instead, it must generate a certain randomly generated backoff period during which it must not request for new reservations.

In order to avoid unnecessary traffic delays during BP merging (as described in Section 4.8.5), a device is allowed to keep its granted reservations (the ones with

Reservation Status bit set to 1) even after it moves its Beacon to the new BP (no need to renegotiate the reservation). The exception is if the reservation conflicts with any BPs, or if it conflicts with other granted reservations in the new BP as long as there are other MASs available to move to. If this last condition does not apply, then the consequence is that the relocating device may still maintain its reservation and the above-described DRP conflict resolution process will have to take care of the conflict.

Modification of reservations can be accomplished very simply. To add MASs to an existing reservation, the owner negotiates a reservation for the new MASs using a new DRP IE. Once accepted by the target, the owner may then combine the two DRP IEs into one, setting the Reason Code to Modified. Once the target recognizes the combination of the DRP IEs, the owner may change the Reason Code to Accepted. To remove MASs from a reservation is even easier. The owner may simply do that with little negotiation. The only thing the owner would have to coordinate with the target is to make sure the target recognizes the changed reservation. To do so, the owner, again, sets its Reason Code to Modified until it sees the target matching its DRP IE.

Termination of a reservation by the owner is as simple as removing its DRP IE from its Beacon payload. The target must follow suit. Termination by a reservation target is done by resetting the Reservation Status bit (back to 0) and giving an appropriate Reason Code. The owner must either follow suit (reset the Reservation Status bit) or simply terminate the reservation. A reservation is also considered terminated if either the owner or the target is not heard (no Beacons or any other communication) for more than three superframes.

Note that all traffic sent during a reservation must be well contained within the reservation block. This includes the mandatory SIFS and Guard Time at the end of the transmissions (see Section 4.6.3).

In order to make sure the medium is being used equitably by all devices, there are certain limitations that are imposed by the MAC specification on the use of DRP reservations. These are described in Sections 4.11.1 and 4.11.2 of MAC Policies.

In conclusion, the DRP management and negotiations represent one of the more complex parts of WiMedia MAC, second only to Beacon management tasks. This is for good reasons, to provide:

- a medium access reservation procedure that is maintained, not by a master, but by peer devices in a neighborhood;
- a fair procedure that is capable of robustly adapting to dynamically changing conditions in a wireless channel without any prior network setup.

This is not a simple feat to achieve, but the WiMedia MAC promises to be the first of its kind to do so.

4.10 PCA

As mentioned before, the main advantage of the WiMedia MAC is its reservation-based medium access capability. That is, instead of letting devices fight each other for access, they schedule their access time through the use of DRP IEs in the Beacon. Consequently, the channel access is made much more efficient, without the need for all sorts of time- and power-consuming coordination and verification efforts on the part of each device. In addition, the use of reservation-based channel access offers a guaranteed QoS to the applications that need it.

Nevertheless, contention-based channel access has its own merits in certain cases as well. This will be clarified later in this section.

The WiMedia MAC also offers the contention-based approach to medium access–PCA. From the MAC perspective, PCA implementation is optional; however, from a PAL perspective, it may be mandatory.

When it comes to PCA, each device contends with its neighbors for access to the medium. In order to avoid collisions, certain access rules have to be followed by all devices. Each device follows the rules in order to achieve a Transmission Opportunity (TXOP).

To give a very quick overview, a device intending to use PCA has to go through the following motions. First, it has to determine if it can use PCA communication in a particular part of the superframe. Then, it must check for channel inactivity. If there is activity on the channel (any received frames with at least a valid HCS), then the device will have to wait until the end of the current transaction. At the end of the transaction, the device will wait an AIFS time (which depends on the user priority; see Section 4.5.3) plus a random number of time periods called Backoff Slots (9 µs each) before it finally achieves TXOP. Now, let us examine the process more closely.

The AIFS time will make sure that the higher priority traffic gets TXOP sooner than the lower priority traffic does. For example, a Voice application has higher priority than the Best Effort traffic, for obvious reasons. The Backoff Slots are there to randomize access attempts by several devices with the same user priority, reducing the probability of collision amongst them. The device maintains a separate Backoff Counter for each user priority it has in order to count down the number of slot periods to its TXOP.

A device that is intending to gain access to the medium via PCA will periodically have to test the channel for activity throughout the process of obtaining

Table 4.9 Access Categories and their priorities

User traffic type	Access Category (AC)	Priority	AIFSn	AIFS (μs)
Voice	AC_VO	Highest	1	19
Video	AC_VI	High	2	28
Best Effort	AC_BE	Low	4	46
Background	AC_BK	Lowest	7	73

TXOP. This is done via the process of clear channel assessment (CCA), which is performed by the PHY. A CCA is performed SIFS time after the end of the last channel-busy period, and every Backoff Slot duration (9 μs) thereafter.

If, during the AIFS period from the last channel activity, a CCA attempt determines that the channel is busy again, then the AIFS period will have to restart. Also, if, during the Backoff Slots, the channel becomes busy again, the Backoff Counters will stop decrementing until the channel is determined clear again. At that time, the countdown can resume after another AIFS period passes without any channel activity detection.

Recall from Section 4.5.2 that the value of AIFS is dependent on the user category of the application data (user traffic). Four user priorities,[21] ACs, are defined in the WiMedia MAC. These are listed in Table 4.9.

The value of AIFS for each AC is then computed as the sum of SIFS duration and the appropriate number (AIFSn) of Backoff Slots:

$$AIFS[AC] = SIFS + AIFSn[AC] \times 9 \; \mu s$$

where the value of AIFSn for each AC as well as the computed AIFS values are given in Table 4.9. The 9 μs in the above equation is the Backoff Slot duration and is the nominal period for the aggregate time to perform CCA, taking into account any associated MAC and PHY processing delays.

Figure 4.28 illustrates the relative AIFS for each AC and gives examples of Backoff Slots for each, resulting in the depicted potential TXOPs for each AC. We emphasize the word potential because there are other criteria to be met in addition to the Backoff countdown. Once the countdown reaches zero, the device achieves TXOP for a particular user traffic only if

- there has been at least an AIFS period of channel inactivity, and
- there is data to transmit, and
- there is no other higher priority user traffic in the device with a zero Backoff Counter value.

---

[21] These are mapped from the eight user priorities defined in IEEE 802.1D specification [4].

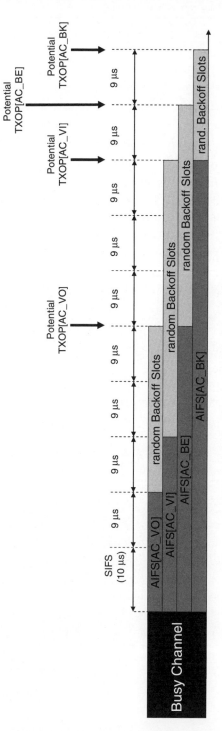

Figure 4.28 AIFS and Backoff Slots for different ACs

Table 4.10   Access Categories' TXOP limits

User traffic type	Access Category (AC)	TXOP limit (MASs)
Voice	AC_VO	1
Video	AC_VI	4
Best Effort	AC_BE	2
Background	AC_BK	2

Once the TXOP is obtained for a given AC, that user traffic must complete within a limited number of MASs. This period is dependent on the AC and is given in Table 4.10. Note that this means, in PCA mode, each device/user may only use the medium for a limited period of time before it has to give up the medium and allow others to contend for it. For voice, this period is short, since voice packets are not expected to contain as much information as the other ACs. On the other hand, video traffic gets the longest TXOP limit of four MAS durations.

For the purposes of PCA communication, a channel is busy not only when the device detects channel activity during the CCA operation, but also when:

- the device is expecting the previous transmission of any of its neighbors to be still ongoing (e.g. for the duration of the frame length as given by the Duration field in the MAC header); or
- the channel is not available for PCA communication (e.g. during BP, or during Hard DRP-reserved MASs, or Soft DRP-reserved MASs of a neighbor when the neighbor is the reservation target and the owner is not a neighbor).

As for the random number of the Backoff Slots, the MAC specifies that a uniformly distributed random integer number generator to be used. The range of this random variable depends on the Access Category. Table 4.11 lists the minimum and maximum ranges for different ACs. Note that both the minimum and maximum ranges increase as the priority of the AC reduces, giving a higher priority AC a higher probability of receiving a TXOP sooner (except for AC_BE and AC_BK, where both get equal probabilities). A Backoff counter is normally started with a random number chosen from the minimum range. However, under certain

Table 4.11   Range of Backoff random variable

Access Category (AC)	Min. Range	Max. range
AC_VO	[0, 3]	[0, 255]
AC_VI	[0, 7]	[0, 511]
AC_BE	[0, 15]	[0, 1023]
AC_BK	[0, 15]	[0, 1023]

circumstances that indicate potential collision (e.g. when the device fails to receive an expected CTS or ACK), the range is systematically increased towards the maximum range.

To clarify the PCA procedure, especially the TXOP determination process, some example scenarios are given below.

Example 1

In this example, we will examine a very simple case of a device trying to achieve a TXOP in the presence of no contention from neighbors. Let us assume the case of a Video application with AC_VI user priority. As per Table 4.9, the value of AIFS for this AC is 28 μs, and according to Table 4.11, the Backoff counter needs to be initialized with a random integer in the range [0, 7]. For this example, let us assume this number is 2. Then, the amount of time the device has to wait (after a busy channel) to obtain a TXOP is, as shown in Figure 4.29, $28 + 18 = 46$ μs ($=$ SIFS $+ 4 \times$ Backoff Slots).

Note, in Figure 4.29, the device is testing channel activity at the beginning of every Backoff Slot boundary using CCA. The fact that no channel activity is detected during each CCA means that the device can proceed with the countdowns uninterrupted during AIFS or Backoff Slots (BS). The figure shows the CCA activity during the two Backoff Slots. Note that the actual CCA operation takes considerably less time than the Backoff Slot period (9 μs) even though it is drawn as if it takes the entire period in Figure 4.29.

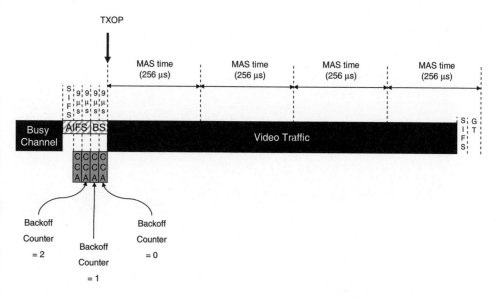

Figure 4.29   Example 1: video user gaining TXOP

Once the device gains a TXOP, as indicated in Table 4.10, it has only four MASs in which to send its frames and still allow for SIFS and Guard Time before the end.

Example 2

This time, let us examine what happens when there is channel activity during AIFS or Backoff Slots. Similar to the scenario of Example 1 above, the MAC sets up the initial AIFS duration as shown in Figure 4.30. This is the same AIFS duration as in Figure 4.29. However, this time, as the device monitors the channel activity every slot, it notices that the channel is no longer idle. (For the sake of simplicity, in this figure the busy channel during the Backoff Slots is shown as taking up only one Backoff Slot (9 μs). In reality, this period could be many multiples of that duration.) Thus, the device waits until the channel goes idle again, after which it restarts its AIFS countdown. After that, the channel stays idle until TXOP, so the process of gaining TXOP is similar to Example 1.

Example 3

This example is a similar scenario to Example 2, but this time we will assume that the random Backoff counter is set to 4. Let us examine the case in which the channel activity is detected during the Backoff Slots. Depending on where in the Backoff duration the 'busy channel' is detected, the device will do one of two things. If the channel becomes busy before the Backoff counter goes to zero, then the counter will freeze. Once the device determines that the channel has been idle for the AIFS duration it may resume the Backoff countdown. This scenario is depicted in Figure 4.31. For the sake of simplicity, in this figure the busy channel during the Backoff Slots is shown as taking up only one Backoff Slot (9 μs). In reality, this period could be many multiples of that duration.

If the channel becomes busy when the Backoff counter is down to zero, the device reinitializes the counter with a new random number and waits for the channel to turn idle and stay idle for the AIFS duration before starting the Backoff countdown.

From the above description and examples of the PCA protocol, it should be quite obvious by now that it is not anywhere near as efficient as the DRP in utilizing the channel capacity. Although necessary to avoid collisions and to allow prioritized traffic to get through sooner, the AIFS and Backoff Slots can dramatically reduce the channel utilization. Hence, it is not expected that PCA will be used by any application other than as a last resort.

One interesting use of the PCA option, however, could be in multicasting command and control frames to a subset of neighbors interested in forming their own subgroup. By scheduling a fixed PCA reservation for the subgroup, all members of the subgroup can participate in those PCA slots; and if there are any application-specific command/control frames to be passed around within the subgroup, then the neighbors can contend for channel access time during those slots. The number

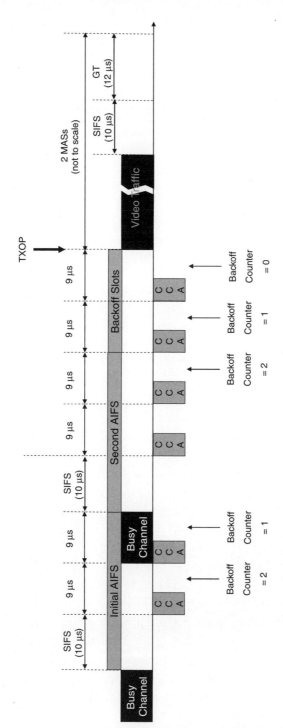

Figure 4.30 Example 2: video user tries to gain TXOP but notices busy channel during AIFS

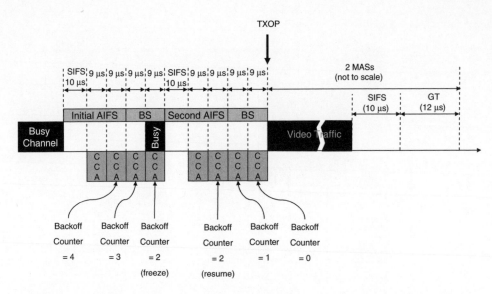

Figure 4.31 Example 3: video user tries to gain TXOP but notices busy channel during Backoff Slots

of slots associated to PCA could be small if such command/control frames are few and far between. The PCA communication relieves such devices from having to set up and maintain group-wide reservations on the fly. Instead, all members of the subgroup will subscribe to the PCA time slots allocated for the subgroup.

4.11 MAC Policies

One of the most condensed parts of the text of the ECMA-368 standard is its MAC policies. This normative annex is only about three pages long, but it will take us many more than that to explain and justify it. It is a mandatory part of the MAC protocol and can take up a fair amount of the processing power of any MAC implementation to conform to it. Thus, it is quite important to understand it.

The philosophy of these policies is as follows. In order to maximize the utility of the channel for the community of the devices, there has to be some sort of 'good-neighbor' mentality among the devices. And when it comes to implementations by multiple competing manufacturers with varying business incentives, such a mentality does not usually come about organically. Instead, it is the job of the standards (and the certification and interoperability programs associated with those standards) to ensure manufacturers are building devices that act in the interest of the community.

Hence, the MAC protocol comes with certain channel usage policies, which put limitations especially on the MAS reservations a device may make relative to those of the other devices in the network. These limitations are described in Annex B of the standard.

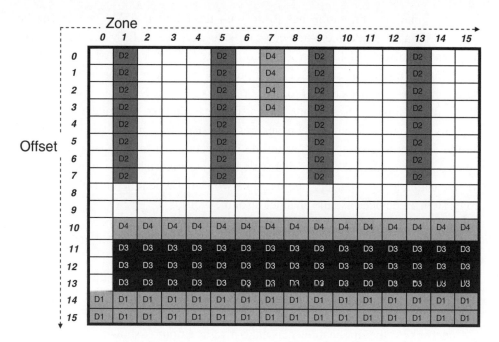

Figure 4.32 Examples of column and row reservations

Another motivation for such limitations to take place is the fact that there are multiple MAC clients that are not interoperable. Consequently, in order for the resources of the MAC not to be frivolously used up by one client at the expense of the other coexisting clients (since there is no coordination possible at the client level), the restrictions are enforced at the MAC level.

Before we start the discussion on the MAC policies, let us define the concepts of a *row reservation* (or row component of a reservation) versus a *column reservation* (or column component of a reservation). The former is any reservation or component thereof that takes the same MAS offsets within every MAS Zone (with the possible exception of Zone 0, where Beacons reside) in the MAS table. The latter is any reservation that is not a row reservation. Examples given in Figure 4.32 clarify them. As shown in the figure, devices D1 and D3 have row reservations and D2 a has column reservation. Device D4 has a row component and a column component in its reservation.

4.11.1 Reservation Size Limitations

The first size limitation in the MAC policies is the declaration that no device may reserve more MASs than it needs for optimal operation. Unfortunately, this is too vague for any certification test to enforce. Nonetheless, it is an expected behavior from the MAC's perspective.

The second size limitation has to do with the number of MASs that can be reserved by a single device. A device can 'safely' reserve 112 MASs (equivalent to 7 out of 16 rows of the MAS table of Figure 4.6). The term *safe* has a very specific meaning in this context, i.e. not preemptable. A device that is *unsafe* may be preempted by other devices to give up the excess MASs it has reserved. The MAC allows unsafe reservations so that devices can take advantage of the full capacity of the superframe whenever possible. However, the MAC expects the unsafe devices to give up on their unsafe reservations (or make them safe) when other devices request them. In this way, a device may take advantage of extra MASs available in a sparsely populated network, but when the network becomes congested, all devices use the channel resources fairly.

The third limitation is the most complicated. It restricts the size of the reservation blocks based on their relative position within the MAS table. The reason for this restriction is that different applications may require different reservation block sizes and service intervals within a superframe for their optimal operation. For example, a video application may require several large blocks of contiguous MASs in a superframe to service its high throughput requirements properly. If left unchecked, the application may take a reservation as shown by the shaded MASs in Figure 4.33. At first glance, there is nothing wrong with this reservation.

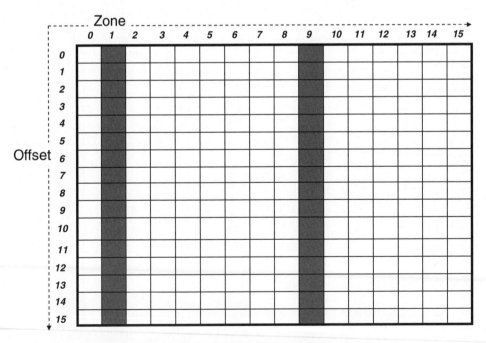

Figure 4.33 Example of an unsafe video application reservation

However, let us now consider a very different application, one in which service intervals have to be very short. The need for a short service interval can be because:

1. The latency has to be short.
2. The memory size of the device is so small that it cannot buffer its application data for too long before it has to transmit them. A Human Interface Device (HID), e.g. a computer mouse, fits this bill.

In this case, a row reservation like the one depicted in Figure 4.34 may be optimal. Now imagine that both of these two applications try to use the medium. Obviously, they cannot coexist since their reservations have a conflict. If only the video application redistributed its MAS allocation to something similar to that of Figure 4.35, then the HID application could coexist with it. The third limitation of the MAC tries to address such scenarios.

The third limitation says that a device may not 'safely' reserve more than X consecutive MASs in a block in the same zone. The largest value is $X = 8$, and that happens if the block starts at the first MAS of any zone (top of a column in the MAS table). However, if the starting MAS of the block is not the first MAS (offset 0) of the zone (perhaps because that MAS is taken by other reservations) then the

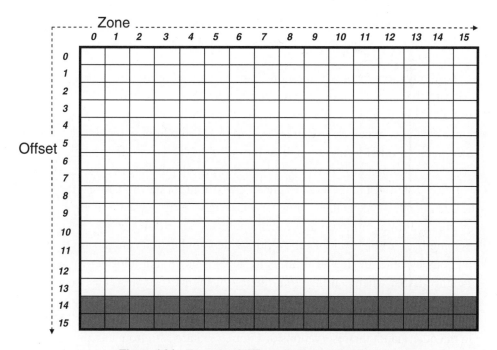

Figure 4.34   Example of HID application reservation

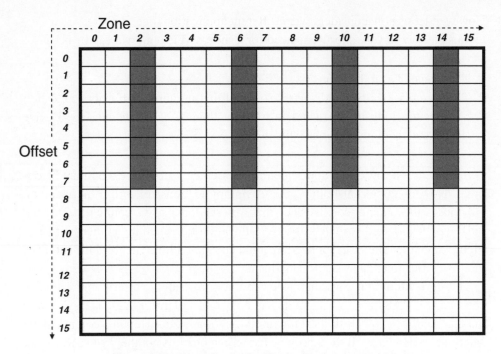

Figure 4.35 Example of safe video application reservation

size of the *safe* block has to be reduced. The reduction is gradual, depending how far the starting MAS of the block is from the first MAS of the zone. The smallest size of the block is 4, when it is taking the last four MASs of the zone),

The motivation for this limitation is to encourage column reservation owners to locate their reservations towards the top half of the MAS table, as much as possible. This then leaves the lower half of the MAS table for row reservations. Of course, as the channel usage increases, it may become impossible to keep column reservations limited to the top half of the MAS table. Then, the limitation allows for safe reservations to encroach into the lower half of the MAS table, but with shorter and shorter block sizes (trying to leave room for new row reservations). On the other hand, if the channel is not very busy, devices may use unsafe reservations (ignore the limitation) to optimize their application performance.

Going back to the example above, the video application can be safe from pre-emptions by other devices if it limits the size of its column blocks of reservation to eight MASs, when starting from the top of the MAS table. Thus, considering Figure 4.33, the reservation is unsafe and the device may be preempted by other devices to give up all MASs it has reserved below the eighth row of the MAS table. On the other hand, the reservations of Figure 4.35 are safe from any preemptions.

The safe/unsafe calculation for each reservation is left to each individual device. This is announced by setting the Unsafe bit to 0 in the DRP IE (see Figure 4.25). All devices that see this bit set to 1 will know that they can preempt the owner of the reservation to give up part or all of its reservation. To do that, the preempting device must send a Relinquish Request IE (see Table 4.3) to the owner or target device of an unsafe reservation. If a target device receives this IE, then it will include it in its Beacon to make the owner aware of such a request. This essentially forwards the Relinquish Request IE from the preempting device to the reservation owner (in case the two are not within a one-hop range of each other).

In the Relinquish Request IE, the preempting device must identify the MASs it is requesting to be relinquished. The owner of the reservation will know that it is the target of this preemption by matching its own DevAddr with the Target DevAddr field in the Relinquish Request IE. Within four superframes of receiving the IE, the reservation owner must either give up the preempted MASs or modify its reservation in such a way that it is no longer unsafe in the preempted DRP IE. Note that, at this point, the device may still carry unsafe DRP IEs, as long as there are no preempted MASs in those DRP IEs.

In order to prevent the back and forth reactions between the preempting device and the reservation owner, there is a hysteresis built into the unsafe reservation mechanism. Certain time limits are put in place as to how fast the owner can reestablish an unsafe reservation after a preemption, and how fast a preempting device can modify its DRP IE to include unsafe MASs. For a device that has just been preempted, the time limit is six superframes unless the preempting device establishes a new DRP IE sooner. For a device that has just preempted a neighbor, the time limit is 32 superframes. That is, the preempting device may not make an unsafe reservation for 32 superframes after the preemption.

4.11.2 Reservation Compaction

One of the most confusing parts of the MAC specification is Section B.4 of Annex B of ECMA-368, where further limitations are placed on the formation of the row and column reservations within the MAS table. Plus, compaction rules are defined here.

First, let us reconsider the justification for such limitations/rules. Recall that, as illustrated by Figures 4.33–4.45, if no thought is given to how each device utilizes the MASs in a superframe, then we could easily get into conditions that certain applications can block other types of applications from channel access, even if plenty of MASs are left in the superframe. The reservation size limitations of Section 4.11.1 are an attempt to prevent such scenarios. Unfortunately, they are not the

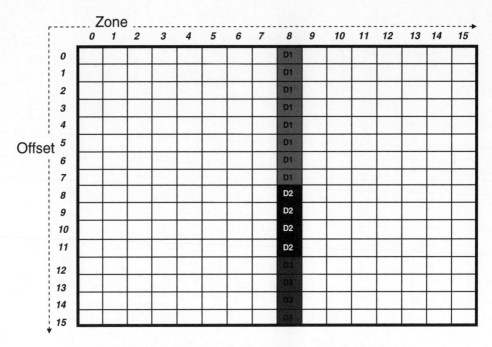

Figure 4.36 Example 1: potentially problematic MAS allocations

only ones needed. Here are two examples of why more restrictions/rules would be necessary to get the best channel utilization from the network.

Example 1

Consider the case of multiple column reservations, each of which is within the required size limits of Section 4.11.1. For example, assume that device D1 needs an eight-MAS column reservation block and that devices D2 and D3 each need a four-MAS column reservation block. By the rules of Section 4.11.1, these devices may place their MAS allocations as shown in Figure 4.36. Note that, although all the devices are limiting the size of their reservation blocks as prescribed, the location of the reservations is troublesome. A new device, D4, needing a very short service interval of about 4 ms (and hence a row reservation) would be excluded from participation in this network, even though there is plenty of capacity left.

Now consider if the same devices were to coordinate their relative MAS locations such that they end up as illustrated in Figure 4.37. Device D4 is now able to take advantage of the capacity of the network.

Example 2

As another example of how coordination of the reservation location can help improve the overall channel utilization in the network, consider a network with a

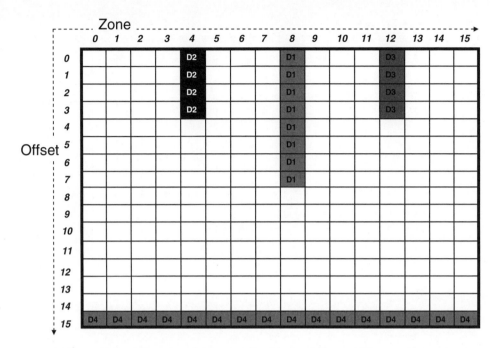

Figure 4.37 A simple reallocation of the reservation blocks of Figure 4.36 can easily enable other devices (D4) to use the network as well

MAS table already occupied by devices D1 and D2 as depicted in Figure 4.38. Note that the BP is also shown in this figure to take up the first eight MASs of the superframe.

Let us now imagine that we have two new devices to join this network: one with a high bandwidth demand and relatively short service interval, D3, and one with a small bandwidth demand and long service interval, D4. For the sake of clarity, let us assume D3 requires roughly eight MASs every other zone (an 8 ms service interval), and D4 only needs eight MASs for the entire superframe. Ideally, one should be able to accommodate these requirements within the remaining channel capacity of Figure 4.38, by fitting the MAS allocations of D3 and D4 as shown in Figure 4.39. Notice that all reservation blocks of D3 are exactly two zones apart, making them about 8 ms apart (32 MASs × 256 μs/MAS = 8.192 ms). No other allocation of unoccupied MASs of Figure 4.38 would satisfy the service interval requirement of D3. This is because the BP is taking up zone 0 of the MAS table, a typical occurrence.

Although the ideal MAS allocations for D3 and D4 as given in Figure 4.39 are simple to determine, due to the lack of a central brain to make such a decision for all devices, the MAC protocol has to rely on each individual device to make the

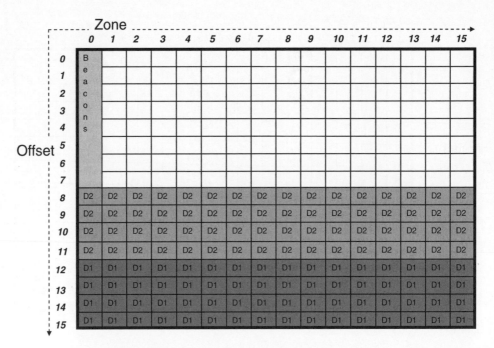

Figure 4.38 Example 2: initial state of MAS allocation

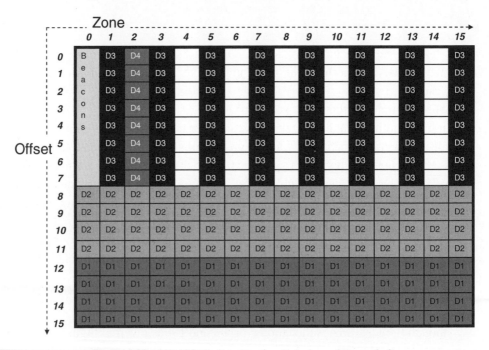

Figure 4.39 Ideal MAS allocation for all devices of Example 2

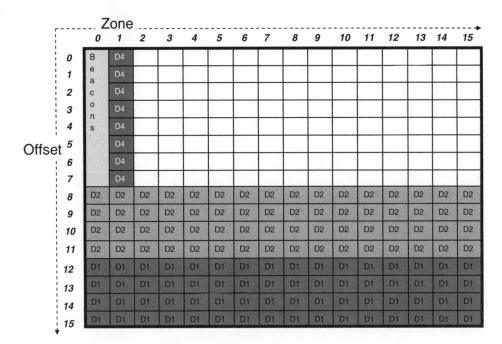

Figure 4.40 An undesirable allocation of MASs by D4 in the absence of D3

correct incremental choice for its own MAS allocations in such a way that is in line with the best interest of the whole community. Plus, depending on the sequence of arrivals (and MAS allocations) of the devices D3 and D4, even with good intentions on the part of each device, there are scenarios in which D3 may be prevented from using the network in the way it needs to. For instance, let us assume D4 arrives first and, in the absence of D3, takes its eight MASs as illustrated in Figure 4.40. The difference in D4's MAS allocations between Figures 4.39 and 4.40 is only the zone number. By taking the zone next to the Beacon zone, D4 is then, in effect, blocking D3 from using the network with its required 8 ms service interval.

D3 can now only take an allocation similar to Figure 4.41, where the irregular zone selection forced by D4's taking up zone 1 is causing D3's service intervals within a superframe to vary from 4 to 12 ms. Hence, the worst-case service interval is too high for D3 to be able to service its application.

The above examples illustrate the need for MAC-level coordination among the distributed devices to place their MAS allocations according to a pattern that would allow maximum superframe utilization. The WiMedia MAC standard came up with the rules listed below as the best compromise between freedom of the individual device to allocate its required MASs and maximizing the overall channel utilization.

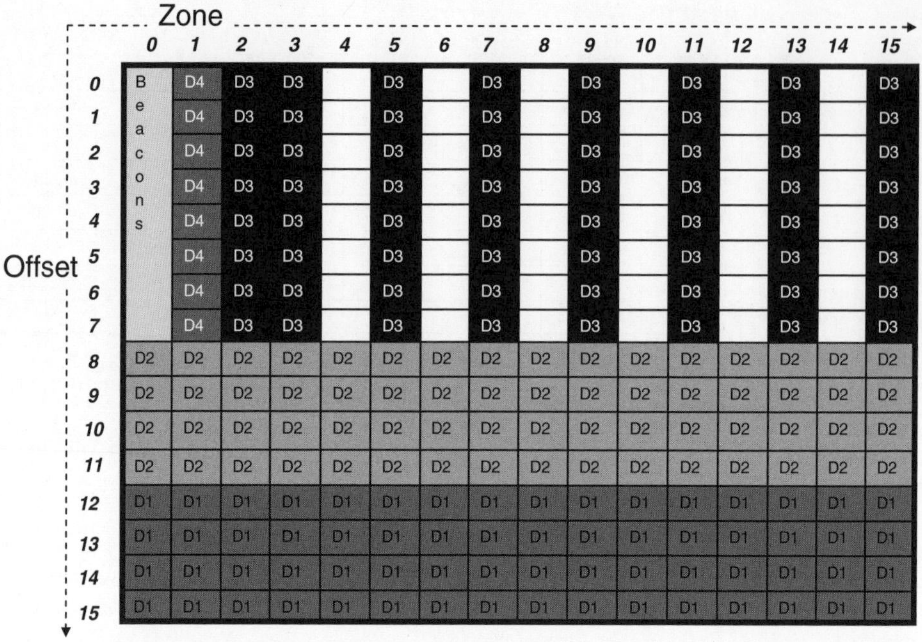

	Zone															
Offset	0	1	2	3	4	5	6	7	8	9	10	11	12	13	14	15
0	Beacons	D4	D3	D3		D3		D3		D3		D3		D3		D3
1		D4	D3	D3		D3		D3		D3		D3		D3		D3
2		D4	D3	D3		D3		D3		D3		D3		D3		D3
3		D4	D3	D3		D3		D3		D3		D3		D3		D3
4		D4	D3	D3		D3		D3		D3		D3		D3		D3
5		D4	D3	D3		D3		D3		D3		D3		D3		D3
6		D4	D3	D3		D3		D3		D3		D3		D3		D3
7		D4	D3	D3		D3		D3		D3		D3		D3		D3
8	D2	D2	D2	D2	D2	D2	D2	D2	D2	D2	D2	D2	D2	D2	D2	D2
9	D2	D2	D2	D2	D2	D2	D2	D2	D2	D2	D2	D2	D2	D2	D2	D2
10	D2	D2	D2	D2	D2	D2	D2	D2	D2	D2	D2	D2	D2	D2	D2	D2
11	D2	D2	D2	D2	D2	D2	D2	D2	D2	D2	D2	D2	D2	D2	D2	D2
12	D1	D1	D1	D1	D1	D1	D1	D1	D1	D1	D1	D1	D1	D1	D1	D1
13	D1	D1	D1	D1	D1	D1	D1	D1	D1	D1	D1	D1	D1	D1	D1	D1
14	D1	D1	D1	D1	D1	D1	D1	D1	D1	D1	D1	D1	D1	D1	D1	D1
15	D1	D1	D1	D1	D1	D1	D1	D1	D1	D1	D1	D1	D1	D1	D1	D1

Figure 4.41   D3's MAS allocations do not meet its application requirement of 8 ms service interval

Of major concern is the dynamics of the allocations over time, and that is where most of the complexity in implementing these rules surfaces. In a static scenario, compact allocations may be made as necessary and with relative ease. However, as devices come and go, or as they modify their MAS allocations over time, the MAS utilization pattern of the superframe can easily change. The originally compact allocations may lose their compact formation over time. This is similar to a highly fragmented hard disk in a computer, which requires frequent defragmentation for optimal performance.[22] Similarly, some level of MAS allocation maintenance is required on the part of each device to keep compaction as tight as feasible. For this reason, the rules in this section are called compaction rules.

In general, the compaction rules try to compact the row reservations towards the bottom of the MAS table and the column reservations towards the top. The compaction rules must be adhered to at all times. At any time, a device has only 32 superframes to adjust its MAS allocations to meet the rules. This is true whether it is in the process of joining the network and negotiating its MAS allocations for the first time, or whether it has had a long-lasting reservation that over time (due to

---

[22] Except that, in a computer environment, optimal performance refers to file access latency, whereas optimality in WiMedia MAC refers to QoS.

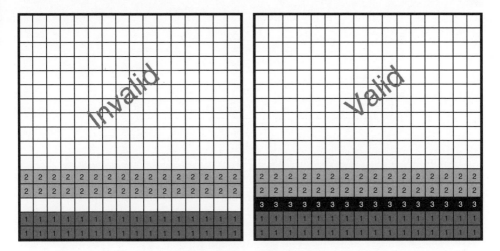

Figure 4.42   Example for Rule 1

other devices leaving the network) is now requiring compaction. The specification does not provide the Safe/Unsafe options of Section 4.11.1 for compaction rules.

The Compaction Rules are as follows.

- *Rule 1*. All row reservation must be as close to the bottom of the MAS table as possible.
  - Exception: owner does not have to break its reservation block into two or more smaller chunks to meet this requirement.

Figure 4.42 illustrates examples of how Rule 1 applies. Notice that, on the left MAS table, device 2 has reservations that are not following this rule. To follow this rule, the device needs to relocate the two rows of MASs down by one row. On the other hand, the right MAS table in that figure has a valid reservation for all the devices shown. Because of device 3 taking the empty row between the reservations of devices 1 and 2, the reservation of device 2 is now valid.

- *Rule 2*. All column reservation blocks must be contained in the top half of the MAS table if possible. If not, then they should be as high up in the MAS table as possible. Also, each reservation block in each zone must be moved up as high in its zone as possible.
  - This rule is subject to the application requirements of the device: service interval and bandwidth. No device can be forced by this rule into allocations inappropriate for its application.

Table 4.12   Zone priorities of Rule 3

Priority	Zone
1	8
2	4 or 12
3	2, 6, 10, or 14
4	1, 3, 5, 7, 9, 11, 13, or 15

The sentences in this rule seem redundant, at first sight. However, there is a very good reason for all of them, and that reason has to do with Rule 3. Rule 2 wants the column reservation users to try to fit their reservations well within the top half of the MAS table whenever feasible. However, it is not requiring that two reservation blocks of, say, four MASs each to be in two different zones. In fact, as we will see in Rule 3, those reservations are expected to stay in one zone after all. Thus, as long as the column reservations do not encroach into the bottom half of the MAS table, the only compaction rule to follow in Rule 2 is to move the reservation blocks up in their zones whenever there is room to do so. However, when the reservation does encroach into the lower half, the highest priority action for the device is to reduce this encroachment as much as possible by relocating the block to a different zone. In fact, this priority is higher than anything Rule 3 requires.

- *Rule 3*. While meeting Rule 2, when an option is available, a column-reservation device must try to allocate its column blocks in zones according to the prioritized list of Table 4.12. Priority 1 means the highest priority and Priority 4 the lowest.

This Rule 3 is very cryptically written in the specification, so let us examine it a bit further. The point of this rule is to address the kinds of issues that were illustrated in Example 2 of this section. Fundamentally, this rule tries to divide the zones of the MAS table into four priority categories. These categories are depicted in Figure 4.43 and are the same as the Priority list of Table 4.12. Notice that, except for the Beacon zone (zone 0), the other 15 zones have been prioritized according to a specific pattern. This pattern is related to the service intervals that can be accommodated.

In order to understand the rationale behind the priority zones, let us start with applications that require column reservations with the shortest service interval that can be accommodated, i.e. 8.192 ms. (Note that a 4.096 ms service interval turns a column reservation into a row reservation.) For these applications to be satisfied, as shown in Example 2 and Figure 4.39 of this section, all of the zones 1, 3, 5, 7, 9, 11, 13, and 15 must be available. This is the same list as in Priority 4 of Table 4.12. Any blockage of any one or more zones in this list by any other device would

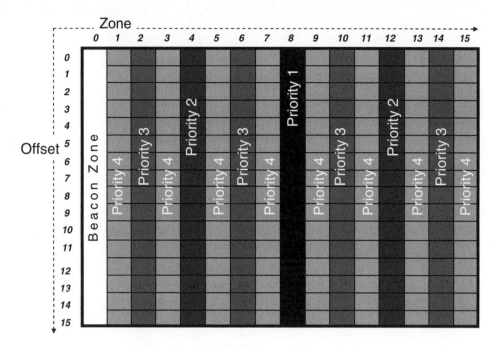

Figure 4.43 Prioritized zones according to Rule 3

cause this application to fail. Consequently, this list of zones should be used by all devices only as the last resort (i.e. only if there are no other zones available, or if the application service interval requires it). This explains the assignment of the lowest priority in Table 4.12.

Next, let us examine applications with service intervals of about 16 ms (16.384 ms to be exact). These applications can be accommodated using allocations in zones 2, 6, 10, and 14 (Priority 3 in Table 4.12). They can also be serviced using zones 1, 5, 9, and 13 (a subset of Priority 4 zones). However, these zones coincide with the zones necessary for applications with an 8 ms service interval. Thus, the preference is to let the 16 ms applications use Priority 3 zones first before they use Priority 4 zones.

Similarly, applications with service interval of 32 ms and 64 ms should use Priority 2 and Priority 1 zones (as defined in Table 4.12) respectively.

Note that we have been using 4, 8, 16, 32, and 64 ms service interval numbers as if each zone has a 4 ms duration. In fact, the numbers do not match exactly. Recall that each MAS is 256 μs in duration. With 16 MASs in each zone, the zone duration computes to be 4096 μs, or 4.1 ms. The entire superframe is also not 64 ms but 256×256 μs = 65 536 μs, or 65.5 ms. Nonetheless, for the sake of simplicity the duration of a zone us usually referenced to be 4 ms.

The MAC policies expect a device to dynamically maintain Rules 1–3 over time. Any time there is a change in the MAS allocations of the neighborhood, the device must re-examine its allocations and reapply this rule within 32 superframes.

Another exception to all the rules in this section is that if a reservation allocation is set as Unsafe in its DRP IE according to the rules of Section 4.11.1, then the compaction rules of this section do not apply. That is, if a reservation is already Unsafe, it need not compact, since it is preemptable at any time. However, as mentioned above, a device may not set a DRP IE as Unsafe as a way of avoiding the rules of this section. There is simply no unsafe mode for compaction rules.

4.11.3 PCA Reservation Rules

The rule on PCA reservations is short. It is designed to address the size limitation issue similar to Section 4.11.1, but specifically for PCA reservations. Recall that PCA communication can only take place during a PCA reservation. However, any device may transmit during a PCA reservation, even in a neighbor's PCA reservation. This makes the ownership of the PCA reservations and limits imposed on its size a bit vague.

To be clear, the MAC specification recommends that every device make a PCA reservation over the MASs it uses even if the same MASs have already been reserved for PCA usage by a neighbor. More importantly, a device is not allowed to transmit in more than 112 MASs in a superframe, unless the Unsafe bit is set to 1 in the DRP IE of the PCA reservation. This limit is consistent with the limit on DRP reservation length of Section 4.11.1.

4.12 Security

As briefly introduced in Section 4.1.11, the WiMedia MAC sublayer offers cryptographically secure communication among devices. It does this by providing data authentication, data encryption, and replay prevention. Authentication (for message integrity) prevents against the man-in-the-middle attack, encryption (for confidentiality) prevents an eavesdropper from deciphering the messages sent in the payloads, and replay prevention offers protection against the replay attack. Note that no association is provided at the MAC level, as this is a MAC client issue. An association provides for a process of one device ensuring the authenticity of another device and exchanging the master keys required for the encryption/authentication engine.

The AES-128 CCM cryptography engine is used for the encryption and authentication purposes. The AES is a symmetric block cipher code which is also adopted by the US Government for the protection of classified information up to the Top Secret level. AES-128 refers to the use of 128-bit blocks. Since it only defines the engine for a single block of 128 bits, the AES algorithm produces a repeatable result given the same input key and data. This is not desirable in cryptography; therefore, the AES engine is usually coupled with a mode of inter-block combining, producing nonrepeating output for the same input. Many different modes of operation have been invented, two of which are called 'Counter,' and 'Cipher Block Chaining Message Authentication Code' (CBC-MAC). Without getting into the details of these modes, a combination of the two gave rise to what is called Counter with CBC-MAC, or CCM. This is the mode in which the AES engine is used in WiMedia MAC, hence the name AES-128 CCM. This mode allows for achieving both authentication (using CBC-MAC mode) and encryption (using Counter mode) features simultaneously. While encrypting the incoming payload, the Massage Integrity Code (MIC) is generated along the way.

The security feature of WiMedia MAC is optional and, hence, it is very tempting to avoid its development into a device. However, since the MAC clients (such as CW-USB) may require secure operation, the decision to omit it from the MAC sublayer needs to be taken very carefully. Unless only a very specialized application is intended for the use of the MAC, it is probably safe to say that the development of the security mechanism in the MAC sublayer is practically mandatory.

Before delving into the security procedures, let us first become familiar with the security fields associated with a secure frame. Recall the general MAC frame format from Figure 4.7. In the Frame Control field, there is a Secure bit that, if set, indicates that the frame payload is secure. Then, the payload (PSDU) format looks as in Figure 4.44. The different security-related fields in this figure are highlighted.

A secure PSDU starts with the security header (the first four fields), followed by the encrypted MSDU, and followed by an MIC. Let us examine each of these fields closely.

The security header starts with the Temporal Key Identifier (TKID). This is the field that holds the temporal key identifier (either Pair-wise or Group) that is used to identify which of the internally stored temporal keys was used to encrypt the MSDU. There will be more on the temporal keys later.

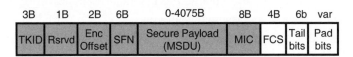

Figure 4.44 Secure PSDU format

The next field to notice in the security header is the Encryption Offset (EO), which indicates the offset from the beginning of the PSDU where the encrypted payload starts. EO allows the use of encryption on all, part, or none of the payload. This is very useful in situations where the payload or part of it needs to be sent unencrypted, while the frame still allows full authentication (uses MIC).

The Secure Frame Number (SFN) is the last field in the security header. It provides for security against the replay attack. It is simply a 48-bit counter that is incremented by the transmitter for every transmitted frame (even on retransmissions). This allows the receiver to ensure that the currently received, authenticated packet is not simply a repeat of a previous one. The SFN also has a function in generating a unique Nonce (one of the input parameters to the AES engine) for every frame.

The MIC field provides for the authentication of the message. Assuming the master key for the device is authentic, the MIC field value, which is output by the AES CCM engine as it is encrypting/decrypting the payload, should match between the transmitter and the receiver. If not, then the message integrity is not authenticated and the packet is likely the work of a 'man in the middle.'

Note that the encrypted portion of the packet is only the secure payload, not the MAC header, the security header, the MIC, nor the FCS fields in the PSDU. However, the authentication of the secure packet takes into account the MAC header in addition to the frame payload.

Now, let us consider the process of establishing a secure connection and subsequently frame-by-frame encryption and authentication. The process starts with the establishment of a shared Master Key. This is a secret 128-bit key that is never transmitted over the air. The MAC sublayer does not define how this master key is established between two devices that trust each other. In fact, this is a topic for the Association Model that is established by the MAC client and layers above. Usually, this is done by way of:

- the use of a cable connection between devices that the user intends to talk to each other;
- the use of numerical verification in both devices by the user;
- the use of NFC technology.

Section 5.1.7, for example, explains the association model for the CW-USB. In any case, for now, let us assume that such a Pre-shared Master Key (PMK) and an associated Master Key Identifier (MKID) have been exchanged between two devices. The MKID is a unique proxy for PMK. Devices may hold multiple Master Keys for different devices. The MKID helps to identify each Master Key to other devices without revealing it over the air. This PMK establishment is only needed once between two devices. Once the devices are introduced to each other using an association model, they should not need to be reintroduced again.

Having a PMK, a pair of devices would then initiate a four-way handshaking sequence, in which they send four PTK Command messages back and forth, establishing shared Pair-wise Temporal Keys (PTKs) based on the PMK as well as other parameters such as the DevAddrs and randomly generated numbers. Similar to the Master Keys, the Temporal Keys are never transmitted over the air. Instead, they are referenced using TKIDs. Once the devices have the shared PTKs established, they will use them temporarily to encrypt and authenticate the frames they are sending/receiving. The duration of validity of the PTKs is equal to the duration of uniqueness of nonces generated by the SFN. Since the SFN is 48 bits long, the nonces will not repeat for 2^{48} frames. This is long enough not to need to update PTK during a communication session. However, when a new session starts between two devices, a new PTK will be established.

Similar to PTKs, Group Temporal Keys (GTKs) are also possible to establish among a group of devices for secure broadcast or multicast communication.

Once a successful four-way handshaking takes place, the two devices will have had a PTK unique to them internally installed (along with their associated TKID), ready for secure communication with each other. When a frame arrives into a device, after checking for validation of the header and payload (HCS and FCS checking) and subsequent acknowledgment as needed, the Secure bit in the MAC header is checked. If it is set, then the frame is supposedly a secure frame. The receiver would then check the TKID in the frame MSDU to make sure it matches one of its installed PTKs (or GTKs). If no match is found, then the frame is to be discarded and the upper layers notified. If a match is found, then the next step is to authenticate the received frame using the installed PTK (or GTK). As mentioned above, authentication is normally a by-product of the decryption process. So, the device may do both authentication and decryption at the same time. If the frame does not authenticate (received MIC does not match the generated MIC), then the frame is to be discarded, whether or not it went through the decryption process. Note that the decryption has to start from the point in the frame payload that is indicated by the EO. The EO is relative to the first byte of the payload. Hence, an EO value equal to the length of the secure payload would mean that that payload is not encrypted at all.

Once the frame is authenticated, it needs to be checked against the replay attack. The replay counter is checked against the SFN field of the frame. If the counter value is the same as or higher than the SFN value, then the frame has to be discarded. Otherwise, the frame is considered secure and the normal MAC-level operations on it can take place and, if decrypted, it can be delivered to the MAC client.

4.13 Ranging

As discussed in Section 3.7, owing to the wideband nature of the WiMedia UWB signals, a very short timing resolution can be achieved. This can be exploited for ranging purposes, and the WiMedia PHY and MAC have optional provisions to allow a device to make accurate time measurements via consecutively exchanged packets between two devices. The time measurement can then be easily translated to distances between the devices.

The PHY takes care of measuring time using counters tied to its high-speed (i.e. high-precision) clock. The minimum theoretical clock rate of the PHY layer is 528 MHz, while practical restrictions require even higher clock rates in most devices. At 528 MHz, the timing resolution is 1.89 ns. A higher clock rate would make this resolution even tighter.

Given the right commands from the MAC sublayer, the PHY would then be able to measure (using its high-speed counter) the elapsed time between two events. Thus, the job of the PHY in ranging is limited to reporting accurate timestamps of transmission/reception events and sending them up to the MAC sublayer for processing.

If a device is able and willing to participate in a range measurement, then it will indicate so in its MAC Capabilities IE in its Beacon. WiMedia MAC defines certain command frame types specifically designed to enable accurate time measurements. These are called Range Measurement command frames (see Table 4.6). The general format of a command frame is given in Figure 4.21. For the Range Measurement command frame the MSDU format is as shown in Figure 4.45. The first byte of this MSDU identifies the type of the Range Measurement frame. There are three such types for now:

- *Type 0: Range Measurement Request*. This type is used for requesting a range measurement from another device (one which has already indicated its range measurement capability in its Beacon). In the payload of this command frame, the initiator of the range measurement includes a 1-byte value indicating how many times the two-way range measurement is to take place consecutively. Using multiple consecutive measurements helps with filtering out the measurement noise.
- *Type 1: Range Measurement*. This type is used for each two-way measurement. There is no measurement payload. It is simply used for time measurements at the receiving device.
- *Type 2: Range Measurement report*. This is the final report sent by the responding device to the initiating device after all the consecutive range measurements are complete. The size of the range payload for this frame depends on the number

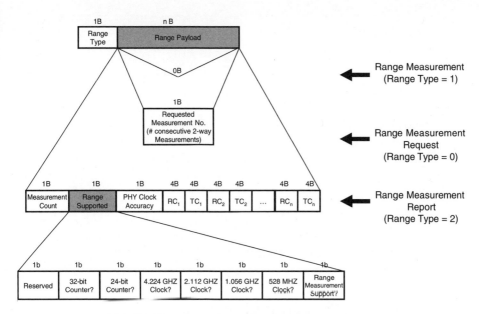

Figure 4.45 MSDU format of Range Measurement Frame

of consecutive measurements made, the Measurement Count. The Range Supported field has a bitmap for each of the range measurement capabilities that the responding device may have. These include counter size (32 or 24 bits) and clock rate (1, 2, 3, or 4 times 528 MHz). The Range Payload also has a field for the PHY clock accuracy in units of parts per million. The rest of the payload is then made up of the pairs of range measurement counter values for each of the two-way time measurements. The pair consists of:

- RC (Receive Count), the counter value when the responding device receives the Range Measurement command frame;
- TC (Transmit Count), the counter value when the responding device transmits the acknowledgment to the Range Measurement command frame.

Now we will examine the process by which the MAC sublayer measures the propagation time between two devices. Referring to Figure 4.46, Device D1 is the initiator and D2 is the responder. D1 initiates the ranging protocol by sending the Range Measurement (RM) command frame to D2, with the ACK policy set to Imm-ACK. D1 records the time of transmission as T1, while D2 records the time of reception as R2. After some elapsed time for the reception of the entire packet and SIFS time, D2 sends an ACK packet to D1, recording T2, the time of transmission. When D1 receives the ACK packet it records R1, the time of reception of the ACK packet.

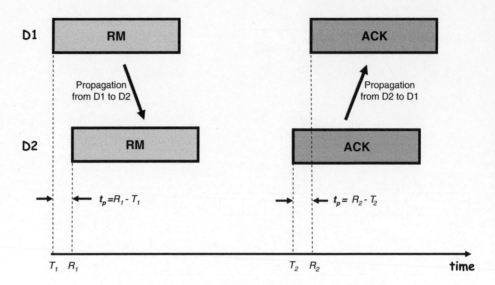

Figure 4.46   TWR using RM command frame

At this point, assuming D1 somehow has access to T1, T2, R1, and R2 times, it can compute the propagation time t_p as an average of the two propagation delays experienced during this two-way exchange between D1 and D2. That is:

$$t_p = \frac{(R_1 - T_1) + (R_2 - T_2)}{2}$$

Note that time is measured in terms of clock cycles in each device's PHY and recorded as counter values in each MAC. Therefore, it is important for D1 to know what the clock speed of D2 is. This is provided in the Ranging Measurement frame type 2 (Ranging Measurement Report).

Since there is room for errors in the time measurement, it is helpful to average out estimation/measurement errors by making multiple measurements. Multiple consecutive ranging measurements may be made by initiating a Range Measurement Request with n consecutive two-way transactions of the RM packet and its corresponding ACK packet. Figure 4.47 depicts the transactions involved among PHY and MAC of each of the two devices for such a scenario.

The initialization step begins with turning on the ranging timer in the PHY and sending a Range Measurement Request from the initiator device (D1) to the responder device (D2). D2 responds with an Imm-ACK, while turning on the ranging timer in its own PHY.

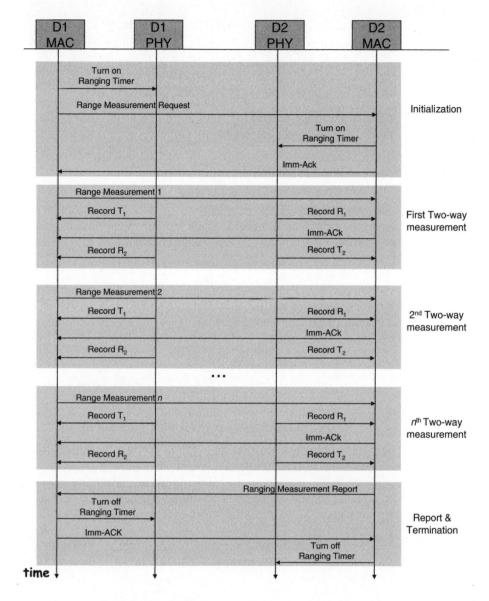

Figure 4.47 Multiple consecutive two-way RM transactions

Then the first TWR transaction takes place by D1 sending the first RM command frame. As explained above, the two devices record their respective timer values T1, R1, T2, and R2. This TWR transaction repeats n times, where n is the number of consecutive two-way measurements that was requested in the Range Measurement Request frame during the initiation step.

At each MAC, the values R1, R2, T1, and T2 are adjusted by the calibration constant of the corresponding PHY to account for its transmit/receive delays.[23] The new counter values are then named RC1, RC2, TC1, and TC2 respectively and used in populating the Range Measurement Report (see Figure 4.45).

Once all the RM transactions are complete, D2 (without any request from D1) sends a Range Measurement Report to D1 indicating all the timer values it recorded along the way, as well as its own ranging parameters, as shown in Figure 4.45. This will allow D1 to compute the average time of flight (propagation time) over all the n transactions as

$$t_p = \sum \frac{RC1^i - TC1^i + RC2^i - TC2^i}{2n}$$

References

[1] Ecma International Standard ECMA-368, 'High Rate Ultra Wideband PHY and MAC Standard,' 2nd edition, http://www.ecma-international.org/publications/standards/Ecma-368.htm, December 2007.

[2] Ecma International Standard ECMA-369, 'MAC-PHY Interface for ECMA-368,' 2nd edition, http://www.ecma-international.org/publications/standards/Ecma-369.htm, December 2007.

[3] 'Guidelines for use of a 48-bit Extended Unique Identifier (EUI-48™),' http://standards.ieee.org/regauth/oui/tutorials/EUI48.html.

[4] '802.1D, IEEE Standard for Local and metropolitan area networks: Media Access Control (MAC) Bridges,' IEEE Computer Society, 2004.

---

[23] T1 and T2 are increased by the calibration constant for D1 and D2 respectively to account for the transmit delays in those PHYs. R1 and R2 are decreased by the calibration constant for D1 and D2 respectively to account for the receive delays in those PHYs.

5

Protocol Adaptation Layer

The Protocol Adaptation Layer (PAL) is considered to be a special layer defined to translate between two protocols. In the case of WiMedia specifications, the expectation has been that this layer would bridge between the MAC sublayer and the existing base of protocols at the upper layers: IP, Bluetooth, etc. More importantly, these PALs would be adhering to the WiMedia MAC policies that ensure a fair and efficient coexistence in the same RF spectrum among applications that are otherwise oblivious of each other. The general idea is that any client of these PALs would be able to make use of the WiMedia Common Radio Platform (MAC and PHY) without having to modify client protocols or applications.[1]

In marketing materials, the WiMedia UWB PALs have traditionally been drawn as in Figure 5.1. As shown in the figure, PALs for USB, Bluetooth, and IP have been conceived so far. However, it is a misnomer that there is a PAL for USB to provide Wireless USB. In reality, the correct drawing should be as given in Figure 5.2. Certified Wireless USB (CW-USB) is not really a PAL, but a full-blown MAC protocol in its own right. It takes over some of the functionalities of the WiMedia MAC. Nevertheless, it adapts/replaces the wired USB by using the majority of the WiMedia MAC functionalities. This will be clearer in Section 5.1.

WiMedia Link Layer Protocol (WLP) ties the WiMedia Common Radio Platform to the Internet Protocol (IP). Unlike CW-USB, WLP builds upon the WiMedia MAC in order to provide IP applications with the services they require. WLP is basically the LLC sublayer of the DLL in the OSI model (refer to Figure 1.13). WLP is the only PAL so far for which the specification is developed fully within the WiMedia Alliance organization.

---

[1] Some modification may be necessary due to the association model or due to the fact that the application will have new capabilities and features to use via the WiMedia platform. These modifications are mostly in the user interfaces and how the user preferences/information/inputs are transferred to the platform.

WiMedia UWB: Technology of Choice for Wireless USB and Bluetooth   Ghobad Heidari
© 2008 John Wiley & Sons, Ltd

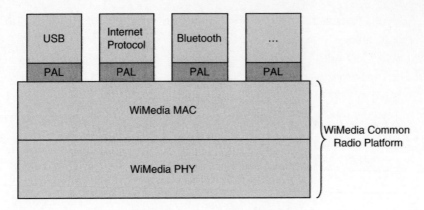

Figure 5.1   Incorrect PAL diagram

The Bluetooth PAL is an adaptation of the WiMedia Common Radio Platform to fit the upper layers of the Bluetooth protocol.

Wireless 1394 was also envisaged to be a WiMedia PAL for IEEE 1394. In May 2004, the 1394 Trade Association approved the development of a PAL for IEEE 1394 over UWB. The wired IEEE 1394 protocol has excellent QoS, making it more suitable for isochronous applications than USB. However, over time the industry lost its interest in short-range WPAN and instead started working on the transmission over coax. As such, no effort is being spent on the IEEE 1394 PAL for WiMedia UWB.

Some general background on CW-USB, High Rate Bluetooth, and WLP was given in Sections 1.3.4, 1.3.4, and 1.3.5 respectively. A detailed description of the different PAL specifications is beyond the scope of this book. Moreover, the Bluetooth PAL and WLP have not yet been finalized or publicly published yet, although

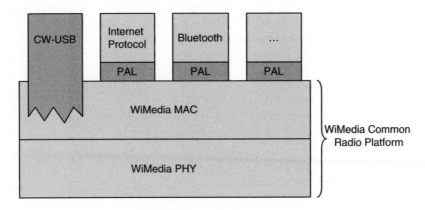

Figure 5.2   Correct PAL diagram

they are close. Thus, at the time of this writing, only the CW-USB specification is publicly available.

Section 5.1 provides a review of CW-USB history, heritage, and protocol. It includes comparisons with USB and WiMedia MAC, as well as the different types of association models and beaconing modes defined within the CW-USB standard. Some implementation issues are discussed at the end.

5.1 CW-USB

From the early days of the MBOA, and later the WiMedia Alliance, providing W-USB was one of the early applications to strive for. The reason was that USB has been such a prevalent bus interface in the PC, consumer, and mobile markets that making it wireless seemed like an easy idea to sell. The use cases are very easy to understand and, with Intel (and so many other prominent companies) behind it, the market for it seemed like the sure thing. Moreover, the rapid adoption of W-USB was expected to be through the introduction of wire adaptors (dongles) that would quickly turn a USB port wireless.

W-USB was on the minds of the competing UWB technologies as well. Freescale, a spin-off of Motorola, was also developing its version of W-USB based on the single-carrier (CDMA) technology of the UWB Forum. Even W-USB solutions based on WiFi and Bluetooth have been developed by others.

Realizing that the name W-USB was no longer unique to the USB-IF (the SIG behind USB and its variations), it decided to qualify its version of W-USB using the new name CW-USB. However, this did not happen until after the first version of the specification was published [1]. In this book, we use the term W-USB when we are referring to the generic idea of a *wireless* USB, whereas the use of the term CW-USB is reserved for the particular version of W-USB that is promoted and standardized by the USB-IF.

The market for W-USB did not develop as fast as it was imagined, mostly because the WiMedia and CW-USB specification (including certification program) development took longer than expected. Nevertheless, the philosophy behind the market friendliness of W-USB is still valid. In fact, compared with WiFi market development in the past, the W-USB market is still very much on track.

In this chapter, we will follow the terminology specified in Section 1.7. In particular, it is worth repeating here that it comes to the words host and device:

- *device* refers to the generic meaning of the word;
- *Device* refers specifically to the USB or CW-USB Device, as it is defined in those protocols (as the slave);

- *host* refers to the meaning of the word in software or firmware context – a host is normally a processor that runs the application on the device;
- *Host* is short for USB or CW-USB Host, which is the 'device' that is the master of ceremony in those protocols.

It is also noteworthy to mention that CW-USB can be considered a MAC in its own right. However, for the purpose of preventing confusing references in this chapter, we will use the word MAC only in reference to the WiMedia MAC.

5.1.1 USB versus CW-USB

Similar to WiMedia UWB, the wired USB specification never became an IEEE standard. Nevertheless, it has been a very successful bus protocol, initially in the PC products sector and later in the CE and mobile communication sectors. Its success is owed to the simplicity and reliability of its usage. A consumer plugs a USB Device into a USB port and, usually without any a priori setup or configuration, achieves the expected connectivity at a very high throughput.[2] The USB Host is usually a personal computer, which is in complete control of the communication link to and from the USB Device.

USB was developed and promoted by the USB-IF, an SIG founded by HP, Intel, LSI, Microsoft, NEC, and NXP. Over the years, the PHY data rate of the USB has grown from a mere 12 Mbps for USB 1.0 to 480 Mbps for USB 2.0. The intent for this bus architecture was to provide a high-speed peripheral connectivity to personal computers. The USB Host has central control over all communication with the USB Devices attached.

The master–slave relationship between a USB Host and a USB Device is the central theme of the architecture of wired and wireless USB. In developing the architecture of CW-USB, the USB-IF borrowed much from the constructs of the original USB architecture. The main objective was to make the upper layers (USB function and client software) as oblivious to the change in going from wired USB to CW-USB as possible.

Of course, the wireless medium is nowhere as reliable and well-behaved as its wired counterpart. In the wired USB, bit error rates of the order of 10^{-11} are common, whereas 10^{-4} is more typical in the wireless world. Moreover, in wired applications, the security and authentication of the Devices on the bus are inherently guaranteed by the USB cable and USB ports. On the other hand, the wireless medium presents all sorts of security challenges, from inability to ascertain the authenticity of the Host/Device, to data interception/eavesdropping and

---

[2] Even though the usage is simple, the protocol is not. Also, high throughput does not mean efficient use of the bandwidth available to the protocol.

Table 5.1 Major differences in wired and wireless channels

	USB	CW-USB
Connection	Only two states: connected or disconnected	Many more sub-states are added to allow for authentication
Secure communication	Inherent: cable connection between the intended Host and Device. No concern for authentication	Wireless channel is inherently insecure. Security requires implementation of proven algorithmic authentication and encryption mechanisms
Inter-packet delays	Short	Long (due to RF transceiver requirements and wireless channel impairments)
Bit error rate	Low ($\sim 10^{-11}$)	High ($\sim 10^{-4}$)
Association	Inherent: cable connection between the USB ports of the intended Host and Device	Multiple association models were developed to determine how a Device and a Host can reliably associate with each other

man-in-the-middle attacks. The wireless channel and the use of RF transceivers also require much larger inter-packet gaps and turnaround times than in their baseband wired cousins. Hence, major changes were made in the bus interface and security to cope with these challenges and to provide the same level of service and security as is expected on the wired side. Table 5.1 lists some of the major differences in going from USB to CW-USB.

In wired USB, the association and authentication of the Device and the Host is inherently made when the user connects the two via the USB cable. In fact, by choosing the ports to plug the cable into, the user manually introduces the trusted Host to the trusted Device. In CW-USB (and W-USB, in general), this is not so simple. The user may have the intention of connecting a Device to a Host, but in reality, due to the broadcast nature of the wireless channel, it could just as easily be connected to another Host nearby or behind a wall. As such, the Host and the Device must first go through an association process to verify that the Host and the Device are in fact the ones the user intended. The association models originally allowed in CW-USB were

- numerical association;
- cable association.

Lately, the USB-IF has agreed to add a third model:

- NFC association.

This model will be introduced in version 1.1 of the CW-USB specification. The association models allow for the exchange of the shared secret key between the

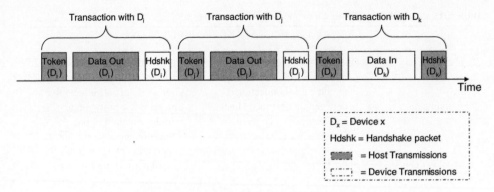

Figure 5.3   Example of wired USB transactions

Host and the Device. This has to be done only once. All future communications between the Host and Device can be based on the same shared secret key.

The CW-USB frame transactions are quite different from their wired counterparts. In USB, each transaction is made up of a Token, a set of data packets (Data In or Data Out), and a Handshake. The Token is sent by the Host, identifying the Device and its endpoint of interest as well as the transfer direction (from Host to Device or Device to Host). The Data In packets are those from Device to Host, and the Data Out packets are from Host to Device.[3] The Handshake packet is an acknowledgement of the reception of data packets. The Host usually cycles through the Devices, giving each one a transaction[4] as often as needed. The Host, being in complete control of the transactions, decides which Device gets a turn next. Each transaction is either 1 ms or 125 µs.

Figure 5.3 illustrates examples of USB Token–Data–Handshake transactions. In this figure, the Host first makes a Data Out transaction with Device D_i, with the Device acknowledging the receipt of that data packet by sending a Handshake packet back to the Host. The second transaction is a similar Out transaction but with Device D_j. The third one, however, is a Data In transaction with Device D_k, where this time the Host sends the Handshake packet as an acknowledgment of the receipt of the packet from the Device.

On the other hand, CW-USB provides for transactions in which many or all Devices get a chance to transmit or receive. Since the turnaround time in the UWB transceivers is relatively long (10 µs), it is best to group several transactions together to allow for the Host to transmit all outgoing packets to various Devices sequentially before requiring it to turn its transceiver around to receive all

---

[3] In USB and CW-USB, the point of reference is always the Host. Hence, an IN direction is from Device to the Host, and an OUT direction is from the Host to the Device.

[4] The term transaction in USB or CW-USB has a slightly different meaning than the one introduced in Chapter 4 for WiMedia MAC.

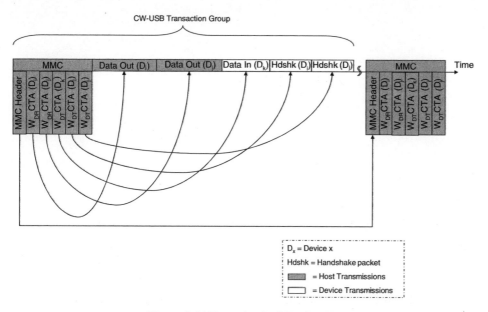

Figure 5.4 Example of a CW-USB TG

incoming packets from various Devices. This would make the protocol more efficient in its transfer throughput.

The Token–Data–Handshake transaction of USB is then transformed into what is called a CW-USB Transaction Group, as illustrated in Figure 5.4. This figure is the exact translation of the example Transactions of Figure 5.3. Instead of the Token Packets, a new type of packet called a Microscheduled Management Command (MMC) is introduced. This packet has pointers to the times in the Transaction Group when each of the incoming and the outgoing packets must start, including Device Handshake packets. The Host Handshake packets are embedded into the MMC packet as well. More on the MMC and Transaction Groups can be found in Section 5.1.3.

Mimicking the USB protocol, there are four types of data transfer defined in the CW-USB protocol: Control, Bulk, Interrupt, and Isochronous. The latter has been modified in CW-USB (relative to USB) to accommodate the unreliability of the wireless channel. Different handshake mechanism and buffering choices had to be added to ensure the isochronous data transfer can be successful in the high error-rate conditions of the wireless medium.

The USB Device classes carry over to the wireless side as they are. However, CW-USB defines an additional class called Wire Adaptor. Depending on whether this adaptor is on the Host side or the Device side, it is more specifically called either Host Wire Adaptor (HWA) or Device Wire Adaptor (DWA) respectively. They allow the CW-USB devices or hosts to attach to an application host via an existing

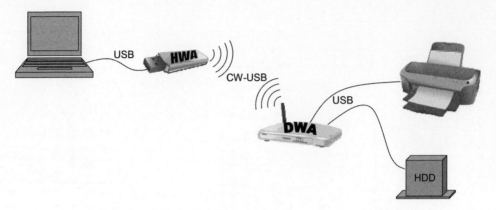

Figure 5.5 Examples of Host and Device Wire Adaptor applications

USB port. See Figure 5.5 for an example of how wire adaptors may be used in the field. In this figure, a printer and a hard disk drive (HDD) are connected to a DWA, while a laptop computer is connected to a HWA, all using their respective USB ports. The HWA and DWA then wirelessly connect the laptop to its peripherals.

One of the advantages of the wireless channel is that it allows for a Host to connect to many Devices without requiring multiple physical connections or ports for each one. In the wired USB, a Host can only reach multiple Devices either through the different ports it has or through the ports of the USB hubs it is connected to. In CW-USB, this limitation is lifted and up to 127 Devices may be addressed by a single Host. This number is merely the number of available addresses for connected Devices. However, as we know well by now from the WiMedia MAC discussions of Section 4.8.1, having 127 Beacons in a Beacon Group is impossible. WiMedia MAC allows at most 96 distinct Beacons to be in a Beacon Group; however, that number is only reachable through the merger of two or more Beacon Groups; otherwise, the Beacon Group may not grow beyond 48 devices. Hence, if we only consider *Self-Beaconing* or *Directed-Beaconing* Devices (no *Non-Beaconing* Devices), then the real maximum number of supportable Devices can only be 48. (These types of Devices will be defined in Section 5.1.4.)

Of course, 48 is still a very large number when it comes to all the Devices connected to a single Host. However, if there are multiple Hosts in close proximity to each other, then they may have to share this maximum number amongst themselves. In that case, each one may end up with fewer than 48 Devices to control.

As in wired USB, a CW-USB Device may have one or more endpoints. Endpoint 0 is by default the control endpoint. This is the endpoint with which the Host communicates during the four-way handshaking and other control-related communications.

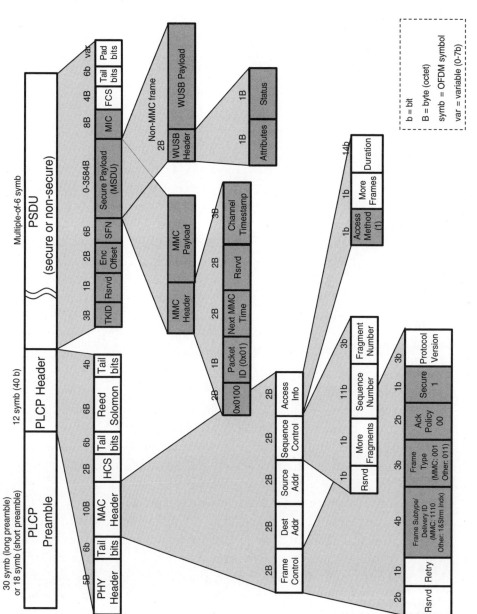

Figure 5.6 General CW-USB frame format

5.1.2 Frame Formats

The CW-USB frame format is encapsulated in the WiMedia MAC frames. As such, the frames follow the general MAC format given in Figure 4.7. Customizing it to the CW-USB frame structure, we get Figure 5.6. Note that the nonsecure payload option is not shown in this figure; however, it is sometimes used. In most circumstances, CW-USB packets are sent via secure MAC frames. (There are exceptions, such as when the Host and Device are not securely 'connected.') The maximum payload size of a CW-USB frame is only 3584 ($= 7 \times 512$) bytes compared with 4075 in a secure WiMedia MAC frame.

The first 2 bytes of the MAC payload contains the WUSB Header, which consists of 1 byte of Attributes and 1 byte of Status. The Attributes field indicates the endpoint number and the packet ID. The Status field indicates the sequence number and status flags about the data or the Device itself. The WUSB Payload field may contain the application data or MMC payload. In the case of isochronous payload, Isochronous Header information is also embedded into the payload field.

Unlike ECMA-368, the CW-USB specification uses the terms packets and frames interchangeably. Hence, what we would normally call a frame in the WiMedia MAC we may refer to as a packet in this chapter. This is done to be consistent with the nomenclature of CW-USB.

There are four types of packet defined in CW-USB: MMC, Data, Handshake, and Device Notification. All of them require encryption, but their EOs are not the same. All except the MMC packet are (a) based on WiMedia MAC's Data frames and (b) require WUSB Header, as shown in Figure 5.6. MMC packets, on the other hand, are based on WiMedia MAC's Application Specific Control frames and, as such, do not have a WUSB Header. Instead, they carry the WUSB Specifier ID number (0x0100) defined in the WiMedia MAC specification.[5] Moreover, since CW-USB has its own retransmission policy, it does not use the WiMedia MAC's acknowledgments. Thus, in all four packet types, the MAC's Ack policy is set to No-Ack.

An MMC packet is the most important packet type in CW-USB. It defines all the transaction management information, including the timing information for each Device to receive/transmit and for the next transaction to start. For this reason, similar to the Beacons, MMCs are sent via the most reliable PHY rate available (53.3 Mbps). The format of this frame type is shown in Figure 5.6 as well. The MMC Header contains information on the CW-USB channel timestamp (which allows all Devices to synchronize to the Host's time), as well as the next MMC

---

[5] ECMA-368 Appendix C

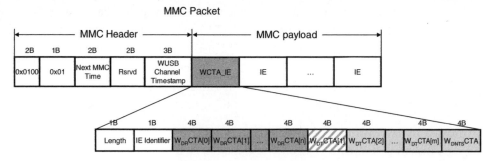

Figure 5.7   MMC packet format

time (which gives the time of the next MMC in microseconds relative to the start of the current MMC preamble).

As shown in Figure 5.7, the MMC payload contains one or more IEs. These tell the Devices what to expect during the transaction. A common IE is the W-USB Channel Time Allocation (WCTA) IE, which contains channel allocation blocks for Device Transmit (DT), Device Receive (DR), and Device Notification Time Slot (DNTS), referred to as $W_{DT}CTA$, $W_{DR}CTA$, and $W_{DNTS}CTA$ respectively. Each block gives the start time by which the Device needs to be ready to execute the appropriate operation (receive or transmit appropriate data, handshake, or notification).

Although an MMC packet is transmitted as a secure packet, the EO field is set to the length of the MMC payload. This way, the entire MMC payload is in plain text (nonencrypted), while the MIC can still be used to authenticate the packet. The plain-text payload is necessary, since Devices need to receive and decode the MMC content even before they establish a secure relationship with a Host.

A Data packet carries the application data for the protocol. It is encapsulated in the WiMedia MAC's Data frame, and is transmitted by the Host or the Device during $W_{DR}CTA$ or $W_{DT}CTA$ time slots respectively. The EO for these packets is set to 2, which renders the WUSB Header fields as plain text.

A Handshake packet is similarly encapsulated in a WiMedia MAC's Data frame. They are only transmitted by the Device during $W_{DT}CTA$ time slots. Similar to MMCs, the Handshake packet is transmitted in plain text using an EO value that covers the entire packet length. These packets are very short and are used for acknowledgment purposes. Similar to Ack frames of WiMedia MAC, Handshake packets are transmitted at the base rate of PHY (53.3 Mbps) to achieve the highest reliability. A Host acknowledges the received packets within the MMC (in the $W_{DR}CTA$ block) and, hence, does not have any need to send Handshake packets.

Device Notification packets are also encapsulated in WiMedia MAC's Data frames. They are transmitted during W_{DNTS}CTA time slots by the Device to make various notifications to the Host. When the Host and Device are not in a secure relationship, these packets may be sent in the nonsecure format; otherwise, similar to MMC packets, they are sent securely, but with EO field value set to the length of the payload, making the payload purely plain text.

5.1.3 Transaction Groups

The sequence of an MMC followed by the time duration defined by that MMC is called a Transaction Group (TG). The sequence of TGs makes a CW-USB Channel. A generic TG is shown in Figure 5.8. Superimposed in this figure is also the structure of a general MMC packet, containing a WCTA IE. The relationship between the various CTA blocks in that IE to the Time Slots in a TG is also depicted. Note that W_{DR}CTA Time Slots always come first after an MMC, followed by W_{DNTS}CTA Slots and W_{DT}CTA Slots. The reason for this particular sequence is to keep the time-consuming RX–TX turnaround times in the Devices and the Hosts to a minimum. The CW-USB Host transmits all that it needs to send to the Devices in a TG before it turns its transceiver around to receive from the Devices. The reception from Devices includes data frames, handshake frames, and Device notifications, if any.

Within DNTS, a number of Device Notification (DN) packet slots (each appropriately sized to a DN packet plus the required guard times) is allocated. This number is adjusted by the Host as needed. Devices may gain access during DNTS using a Slotted Aloha protocol. Since all channel allocations are made by the Host, if a Device needs to get the Host's attention for any reason, then it will use DNTS to send DN packets. This is especially necessary when new Devices need to associate with the Host.

Figure 5.8 also shows that each W_xCTA block (where x stands for DR, DT, or DNTS) points to the start of a specific Time Slot in the TG. By decoding the MMC, Devices determine at which time during the TG to receive, to transmit, or to attempt notifying the Host.

Figures 5.9 and 5.10 illustrate the relationship of CW-USB TGs to WiMedia MAC's superframe. Figure 5.9 shows the one-dimensional version; Figure 5.10 depicts the same, but in a 2D space (MAS table). Each Transaction is fit into a block of MASs that the Host has previously reserved (as a WiMedia device) as part of a row reservation using DRP IEs in the Host's Beacon frame. Within each TG, there is an MMC followed by OUT data time slots, then DNTS and IN data/handshake time slots. The duration of each of these slots may change from TG to TG.

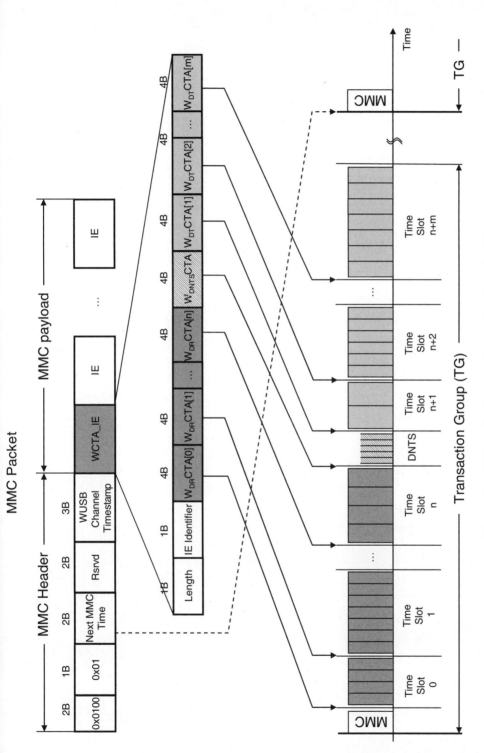

Figure 5.8 MMC packet format and its relationship to the TG

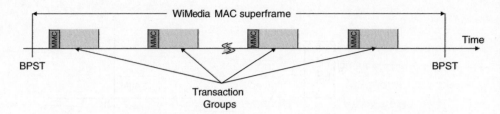

Figure 5.9   CW-USB TGs within a WiMedia MAC superframe (one-dimensional view)

5.1.4 Device Beaconing

The CW_USB specifications allow for three types of Device: *Self-Beaconing*, *Directed Beaconing*, and *Non-Beaconing*. The Self-Beaconing Device (SBD) acts independently from the CW-USB Host in terms of fulfilling its Beacon protocol responsibilities. It maintains its own Beacon management, including BPST, BP, Beacon IEs, etc. It also performs at least some basic DRP reservations and management. A Self-Beaconing type of Device is necessary if a DRD, defined in Section 5.1.6, is to be implemented. Also, if a device with multiple Host connections or multiple PAL implementations is intended, an SBD is a necessity.

A Directed-Beaconing Device is one that has no knowledge of the Beaconing protocol of the WiMedia MAC. The Beaconing operation of this Device is

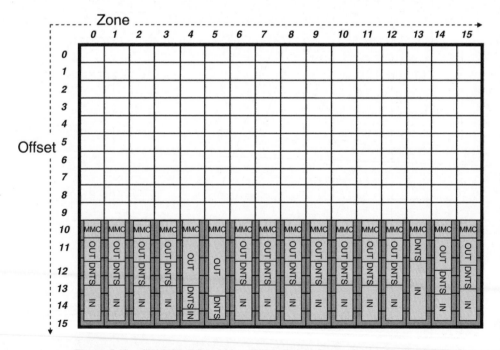

Figure 5.10   CW-USB TGs within a WiMedia MAC superframe (2D view)

remotely controlled by the Host. The Host instructs the Device through CW-USB commands when and what packet to send during the BP. Also, the Device is instructed to receive the Beacon packets during the BP and ship them to the Host for analysis. This way, much of the implementation complexities of Beacon management are removed from the Device, potentially making it a less expensive device. However, the concept of remote or directed Beaconing gets into trouble when it comes to WiMedia MAC's strict response time requirements. More importantly, as a device operating under the jurisdiction of the WiMedia MAC, a CW-USB Device's responsibility is to adhere to all the WiMedia MAC requirements, regardless of where the processing brain resides. According to the MAC protocol, before a Device communicates with a Host (e.g. to connect to the Host for the first time), the Device is required to join the network by first listening to the BP, finding an empty Beacon Slot, and then sending a Beacon in the BP to announce its presence to the other devices. However, for the Directed-Beaconing Device to be able to transmit a Beacon in the correct time slot, it first has to communicate with the Host and receive instructions on what packets (Beacons) to receive, transmit those to the Host, then receive further instructions on how to construct the Beacon and when to send it. Hence, technically speaking, Directed Beaconing violates the WiMedia MAC protocol for joining a network.

Taking the Device simplification even further, the Non-Beaconing CW-USB Device is conceived as one that does not require any Beacon management, nor does it have to send or receive Beacons. Such a Device, having a much lower TX power, is considered to be deployed so close to the CW-USB Host that the Device's coverage area is completely within that of the Host. As such, the Host receives and transmits all the Beacons that otherwise the Non-Beaconing Device would need to. The motivation for this Device is to decrease the implementation complexities of the Device as much as possible. However, as the wireless channel is an unpredictable environment (in coverage), such an approach is very risky and has to be carefully considered.

In the rest of this chapter, we will focus mostly on the SBDs.

5.1.5 Data Flow and Connection Process

The Host decides the data rate for all incoming and outgoing data packets, provided the Device can support it. The incoming packet data rates may change every TG. As mentioned before, acknowledgments are made without the use of WiMedia MAC's acknowledgment mechanisms (set to No-ACK policy). The incoming packets (from Device to Host) are acknowledged by the Host within the W_xCTA blocks. However, the outgoing packets (from Host to Device) are acknowledged through the use of handshake packets during the handshake time slots. Retransmission may occur if acknowledgments are not received.

The typical sequence of operation of a Self-Beaconing CW-USB Device can be summarized as follows. Let us assume that the Device and the Host are associated and have established a Connection Context with each other by way of one of the association models (see Section 4.12). Upon initial power-up, the Device is in an *unconnected* state. After finding a Beacon Group, it starts to search for the Host of interest by scanning for MMCs within the superframe to look for the Connection Host ID (CHID) of interest. If no MMC is received (or no MMC from the Host of interest is received), then the Device may choose to change channel and repeat the search. Once an MMC from the CHID of interest is received, the Device attempts connecting to it using one of the DNTSs.

However, as with any other WiMedia Device, the Device must first adhere to WiMedia MAC's protocol of joining the Beacon Group and reserving MASs. The Device sends Beacons that include a private DRP IE that covers the same MASs reserved by the Host, if available. If not completely available, the Device reserves as much as is available.

Once established within the WiMedia neighborhood, the Device uses the default unconnected CW-USB Device Address of 255 for sending DN_Connect notifications to the Host during DNTS. Once the Host recognizes that the Device is intending to connect, it acknowledges the connection notification and assigns it an unauthenticated Device Address in the range 128–254.

At this point, the Device is connected but unauthenticated. It will then listen to the Host, which will allocate time slots for the Device to proceed with the four-way security handshake protocol to obtain a session key. Once complete, the Device is assigned an address in the range 1–127 and is considered *connected*.

5.1.6 Dual Role Device

Similar to the USB OTG [2], which provides for a device to have both a USB Device functionality and a limited USB Host capability, CW-USB's DRD refers to a device that can act as both a CW-USB Device and a limited CW-USB Host. This is useful in providing certain computer peripherals (usually Devices) the capability to also act as Hosts to some other peripherals. Moreover, as portable devices are becoming more popular, the DRD can play an important role in allowing consumer electronic and mobile communication devices to become connected, without the intervention of a computer.

Figure 5.11 illustrates several scenarios in which a DRD would be useful.

Even though the DRD is defined using the concept of a 'limited Host,' there is no formal specification on what a limited Host means. In fact, the whole DRD

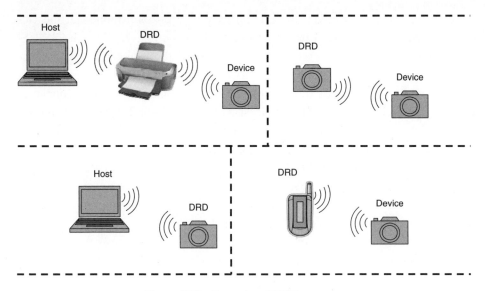

Figure 5.11   Examples of DRD use cases

definition in the CW-USB specification is quite short and ambiguous.[6] Much of it is left to implementation. Clearly, as the DRD popularity increases in the market-place, the specification will have to become more detailed.

For now, the specification defines two modes of operation: Combination DRD and Point-to-Point (P2P) DRD. The former refers to the capability of a DRD independently and simultaneously acting as a Device to a Host and as a limited Host to another Device. A printer, as illustrated in Figure 5.11, is a good example of such a mode of operation. On the other hand, the P2P DRD is in relation to a special case where two DRDs act as both Host and Device to each other, creating a two-way link only between themselves.

There are other scenarios that can be envisaged for DRDs, which are not explicitly defined in the standard. For example, there is a case to be made for a DRD to act as a Device or as a limited Host, but not simultaneously. That is, depending on circumstances, a product can decide to become a Device or a Host to communicate with another Host or Device. Although this does not offer more than the Combination DRD does, it certainly can reduce the DRD complexity and cost.

5.1.7 Association Model

The Association Model [3] is the supplemental specification to the CW-USB standard on how a Host and Device may establish a first-time, trusted connection.

---

[6] The entire section on DRD in the CW-USB specification is only 1.5 pages long.

Without an association model there is no easy way of confirming that the Host or the Device is in fact what it purports to be. Imagine that you have a CW-USB Device peripheral, such as an external hard disk, that you would like to connect wirelessly to a CW_USB Host computer. The hard disk user interface may show that it detects a Host and you may assume it is the one that you intend to connect to. In reality, though, owing to the broadcast nature of the wireless channel, it could be the computer of a neighbor or, worse, that of an eavesdropper across the street using a high-gain antenna to pretend to be a nearby Host.

This problem does not exist in the wired USB world, since the user manually makes the secure association between the Host and the Device by connecting the USB port of one to that of the other, thereby ensuring that there is no mistake or foul play. Furthermore, the cable between the ports acts as the means by which security of the data transmission is ensured.

The encryption and authentication features provided in the MAC protocol assume that there is a secret key shared between the Host and the Device. However, the WiMedia MAC does not concern itself about how this secret key is shared in the first place. This is the job of the MAC client, in this case CW-USB.

Although there are mechanisms to exchange secret keys over the air, such wireless data exchange cannot be protected against man-in-the-middle attacks unless there is an initial trusted association between the Host and the Device. The Association Model of version 1.0 of the CW-USB standard prescribes two association models:

- Cable Association;
- Numerical Association.

The Cable Association Model is simply based on the use of a USB cable connection between the Host and the Device. The user does this once to associate the two together. Through this connection, the Host queries the Device as to whether it has a previous Connection Context with this Host. If not, then it establishes one with the Device; otherwise, the Host will refresh it with a new Connection Key (shared secret key).

The Numerical Association Model is the second method defined by the Association Model specification. It is based on the Diffie–Hellman protocol, which is a method of establishing a shared secret key between two parties over an insecure medium. In this case, the protocol allows the shared secret key to be established between the Host and the Device over the air. Since the Diffie–Hellman protocol is susceptible to man-in-the-middle attacks, the model requires some sort of user intervention to prove that the intended Host and Device are, in fact, talking to each other. This intervention is made by requiring the user to verify that the numeric values displayed on both the Host and the Device are identical. The same numerical value is generated at the Host and the Device using the shared secret key established by the Diffie–Hellamn protocol. Once the numerical validation is made, the Host establishes the Connection Context with the Device, completing the association process. Even though the Numerical Association model is sometimes called the 'in-band' model, referring to the fact that the Connection Key (shared secret key) is established over the same medium (the air) as the subsequent data communication, there is still an Out-of-Band element to it: the user validation of the numerical values.

Hosts must be able to do both Cable Association and Numerical Association (four digits). Devices can do either one, but they have to do Cable Association if they have a USB port. If they have a display, then they have to implement Numerical Association (at least two digits). If they have both a USB port and a display, then they have to implement both models. For DRDs and embedded Hosts, the requirements may be a bit more relaxed compared with personal computers, depending on a number of factors. However, in general, it is safe to assume that the Hosts must be able to support any type of Device association necessary to allow Devices to associate with them.

There is another association model that is planned to be added to the list of approved models for CW-USB in the next revision of the specification. This one is based on the novel NFC technology. The idea is that the Host and the Device are brought to very close proximity of each other (less than 10 cm), which enables NFC to take place. The special property of NFC that allows communication to take place only in close proximities provides for the protection against man-in-the-middle attacks.

The intent of the association model is to make the task of association a one-time affair. After a first connection, the Host and the Device may recall the Connection Context as many times in the future as necessary. Provided neither one loses or resets this context, they do not have to associate with each other ever again.

5.1.8 Implementation and Interface Issues

In designing a CW-USB product, there are many high-level questions that need to be answered:

- Is it a Device, a Host, or a DRD?
- Is it a wire adaptor or a native Device?
- If Device, is it Self-Beaconing, Directed Beaconing, Non-Beaconing?
- What type of association model to use?
- What is the interface to the application host?

- Will it be a single-chip solution or a chipset?
- What band groups will the RF frontend cover?
- Etc.

Each of these choices will have a dramatic effect on the overall architecture of the end solution and needs to be investigated carefully.

The implementation will especially depend on whether the product is destined to be a CW-USB Host or a Device. A fully fledged CW-USB Host requires being a complete WiMedia device. This, in turn, means that the entire WiMedia MAC has to be implemented. A simple CW-USB Device need not, however. For instance, MAC features such as Immediate Ack, Block Ack, retransmissions, and PCA, among others, can be omitted form a CW-USB Device. On the other hand, for a CW-USB Device (or Host), the implementation of the security features of the WiMedia MAC is no longer optional.

The usual expectation is that a CW-USB Device is much simpler in implementation than the Host, since the Host takes on most of the protocol management burden. However, there are areas in which a Device is more complicated than a Host. For example, since the Device does not have any control over the arrangement of W_xCTA Time Slots in a TG, then it has to be ready for the worst-case scenario. That scenario would take place when the Host schedules a W_{DT}CTA slot right after the MMC (no OUT packets other than MMC). This means the Device has to receive the MMC and in the time between the end of the MMC packet and the beginning of the first CTA (10 µs), decode the MMC, turn around its PHY from Receive to Transmit, and start transmitting the packet.

The question of whether the product will have other PALs (such as Bluetooth or WLP) included will also affect the decision of whether to implement the full WiMedia MAC and whether what interface between MAC and CW-USB would be more reusable for the other PALs.

In addition to the above choices, designing all the different parts of a W-USB chip or chipset may require expertise that is not always present in one company. In that case, outsourcing or licensing intellectual property, or even procuring part(s) of a chipset, may be necessary. If so, modular design of the architecture will greatly help the full integration of the system down the road. A modular approach is very helpful in being able to unit-test each module to its full specification before trying to test the bigger system. This will allow debugging activity to proceed much faster and in parallel among all modules. The modular approach may be as coarse as dividing the system into RF, Baseband, MAC and CW-USB, interface to the application host, and host drivers. On the other hand, the modularity could be in finer details within each of these coarse modules.

Figure 5.12 Abstract representation of the relationship between CW-USB and WiMedia MAC

As mentioned before, among all WiMedia PALs, CW-USB does not quite fit the framework of a stand-alone PAL. The CW-USB protocol acts much like a MAC sublayer on its own, while, at the same time, getting support from the WiMedia MAC. In fact, as represented in Figure 5.12, the CW-USB protocol delves into the WiMedia MAC sublayer, replacing some of the WiMedia MAC functionalities (while using others), to complete the control of the medium access. As such, there is no standard interface between CW-USB-embedded firmware and the WiMedia MAC. The combination of CW-USB and MAC has to be somewhat developed jointly for them to be able to integrate properly.

If all of the modules are developed in-house, then proprietary interfaces among them can easily be defined in-house just as well. However, if some of the modules are expected to be outsourced or procured, then the necessity for a common or standard interface will become quite apparent. This is the main reason, for example, why the MPI was developed and published by the WiMedia Alliance (ECMA-369 standard specification [4]), even though it is not a mandatory standard. This way, PHY developers can easily integrate their solutions with MAC developers.

The need for such standard interfaces is also apparent in the upper layers. For example, the interface between the CW-USB radio module and the computer host has been developed based on both Peripheral Component Interconnect (PCI) and USB transports. For the USB transport, the interface is called the Radio Controller Interface (RCI) and is defined as part of the W-USB specification [1] (along with the two associated errata documents [5, 6]). For the PCI transport, this interface is called the Wireless Host Controller Interface (WHCI) [7], being developed by Intel. These interfaces are only applicable to CW-USB Host development (connecting via PCI or USB). On the Device side, the use of Secure Digital Input/Output (SDIO), USB, or a proprietary solution such as memory-mapping, among others, may be reasonable, depending on the Device application.

Work is still in progress to create a harmonized interface scheme that can apply to multiple PALs and scenarios. However, depending on whether the solution is for a personal computer environment or for portable CE, the approach could vary dramatically.

In a personal computer environment (CW-USB Host), the emphasis is on letting the personal computer host (operating system and the associated drivers) do most of the work of controlling the radio. In this case, the driver is completely aware of the WiMedia MAC protocol. As mentioned, WHCI is being developed for the special case of PCI-based transport between CW-USB Host Controller (Wireless Host Controller (WHC)) on the radio module and the CW-USB Host Controller Driver (Wireless Host Controller Driver (WHCD)) running on the personal computer. As part of the WHCI specification, an interface between the UWB Radio Controller Driver (URCD) and the UWB Radio Controller (URC) is also defined, entitled UWB Radio Controller Interface (URCI). Figure 5.13 illustrates WHCI and URCI. Logically, similar to the RCI interface, URCI allows the personal computer to control the URC for WiMedia Radio management functionalities, including Beacon

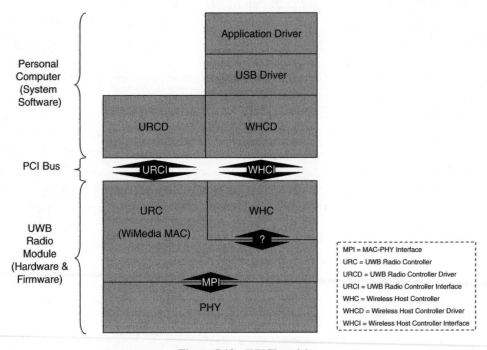

Figure 5.13   WHCI model

and DRP management. At the same time, WHCD can control the W-USB Host functionalities of WHC through the WHCI. Note that there is no definition of an interface between the WHC and WiMedia MAC.

As mentioned before, the WHCI model assumes that much of the control resides on the system software side. As such, the application host processor takes much of the burden for running the CW-USB module. It is extensible to allow other PALs to use a similar interface model on a personal computer.

The WHCI model does not necessarily fit the requirements of the embedded products, however, especially portable ones, such as mobile phones and cameras. Such products usually do not have a very powerful processor and/or excess processor capacity to run radio control functionalities. Instead, the interest is in reducing the burden on the host processor to the minimum achievable, allowing each radio module to be as independent as possible and only to engage the host processor when needed. The host processor is usually radio agnostic. For example, the next generation of Bluetooth, i.e. High Rate Bluetooth, requires such an interface approach for implementation in, say, a mobile phone. This interface is not yet defined, but will likely be done in the near future.

References

[1] Agere Systems, Inc., *et al.*, 'Wireless Universal Serial Bus Specification,' revision 1.0, May 12, 2005.
[2] USB Implementers Forum, 'USB On-the-Go,' http://www.usb.org/developers/onthego.
[3] Ecma International Standard ECMA-369, 'MAC–PHY Interface for ECMA-368,' 2nd edition, http://www.ecma-international.org/publications/standards/Ecma-369.htm, December 2007.
[4] Agere Systems, Inc., *et al.*, 'Association Models Supplement to the Certified Wireless Universal Serial Bus Specification,' revision 1.0, March 2, 2006.
[5] Agere Systems, Inc., *et al.*, 'Wireless Universal Serial Bus Specification Errata on Revision 1.0 as of July, 2005,' July 2005.
[6] Agere Systems, Inc., *et al.*, 'Wireless Universal Serial Bus Specification Errata on Revision 1.0 as of December,' December 2006.
[7] Intel Corporation, 'Wireless Host Controller Interface Specification for Certified Wireless Universal Serial Bus,' revision 0.95, June 16, 2006.

Appendix A

Acronyms

2D	two-dimensional
3G	third-generation
4G	fourth-generation
A/D	Analog-to-Digital Converter
AC	Access Category
ACK	Acknowledgment
ADC	Analog-to-Digital Converter
AES	Advanced Encryption Standard
AGC	Automatic Gain Control
AIFS	Arbitration Inter-Frame Space
AMP	Alternate MAC–PHY
ARQ	Automatic Repeat-Request
ASIE	Application-Specific Information Element
AWGN	Additive White Gaussian Noise
B-ACK	Block Acknowledgment
BcstAddr	Broadcast Device Address
BM	Burst Mode
BP	Beacon Period
BPOIE	Beacon Period Occupancy Information Element
BPSK	Binary Phase Shift Keying
BPST	Beacon Period Start Time
CBC-MAC	Cipher Block Chaining-Message Authentication Code
CCA	Clear Channel Assessment
CCM	Counter mode CBC-MAC (Cipher Block Chaining Message Authentication Code)

CDMA	Code Division Multiple Access
CE	Consumer Electronics
CEPT	Conference of Postal and Telecommunications Administrations
CHID	Connection Host ID
CM	Channel Model
CRC	Cyclic Redundancy Check
CRTC	Canadian Radio-television and Telecommunications Commission
CSMA	Carrier Sense Multiple Access
CSMA/CA	Carrier Sense Multiple Access with Collision Avoidance
CTS	Clear To Send
CW-USB	Certified Wireless Universal Serial Bus
D/A	Digital to Analog (converter)
DAA	Detection-And-Avoidance
DAC	Digital-to-Analog Converter
DBD	Directed-Beaconing Device
DCM	Dual Carrier Modulation
DevAddr	Device Address
DFT	Discrete Fourier Transform
DLL	Data Link Layer
DME	Device Management Entity
DN	Device Notification
DNTS	Device Notification Time Slot
DR	Device Receive
DRD	Dual-Role Device
DRP	Distributed Reservation Protocol
DS-CDMA	Direct Sequence CDMA
DSP	Digital Signal Processing
DSSS	Direct-Sequence Spread-Spectrum
DT	Device Transmit
DWA	Device Wire Adaptor
EC	European Communities
ECC	European Communities Commission
EIRP	Equivalent Isotropic Radiated Power
EO	Encryption Offset
ETSI	European Telecommunications Standards Institute
EU	European Union
EUI	Extended Unique Identifier
EVM	Error Vector Magnitude
FCC	Federal Communications Commission
FCS	Frame Check Sequence

FDS	Frequency-Domain Spreading
FEC	Forward Error Correction
FER	Frame Error Rate
FFI	Fixed-Frequency Interleaved
FFT	Fast Fourier Transform
GPS	Global Positioning System
GTK	Group Temporal Key
HCS	Header Check Sequence
HDMI	High-Definition Multimedia Interface
HID	Human Interface Device
HWA	Host Wire Adaptor
IC	integrated circuit
ID	Identifier
IDFT	Inverse DFT
IE	Information Element
IFFT	Inverse Fast Fourier Transform
IFS	Inter-Frame Space
Imm-ACK	Immediate Acknowledgment
IP	Internet Protocol
ISI	Intersymbol Interference
ISO	International Organization for Standardization
ITU	International Telecommunication Union
KCK	Key Confirmation Key
LLC	Link Layer Control
LQE	link quality estimate
LQI	Link Quality Indicator
lsb	least-significant bit
MAC	Medium Access Control
MAS	MAC Access Slot
MBOA	MB-OFDM Alliance
MB-OFDM	Multi-Band OFDM
MCDU	MAC Command Data Unit
McstAddr	Multicast Device Address
MIB	Management Information Block
MIC	Ministry of Internal Affairs and Communications (Japan)
MIC	Ministry of Information and Communication (Korea)
MIC	Message Integrity Code
MIFS	Minimum Inter-Frame Spacing
MII	Ministry of Information and Industry
MIMO	Multiple-Input Multiple-Output

MKID	Master Key Identifier
MLME	MAC sub-Layer Management Entity
MLS	Microwave Landing System
MMC	Microscheduled Management Command
MPDU	MAC Protocol Data Unit
MPI	MAC–PHY Interface
msb	most-significant bit
MSDU	MAC Service Data Unit
NAV	Network Allocation Vector
NBD	Non-Beaconing Device
NFC	Near Field Communication
No-ACK	No Acknowledgement
Ofcom	Office of Communications
OFDM	Orthogonal Frequency Division Multiplexing
OSI	Open Systems Interconnection
OTG	On-The-Go
OUI	Organizationally Unique Identifier
P2P	Point-to-Point
PAL	Protocol Adaptation Layer
PAN	Personal Area Network
PAR	Project Authorization Request
PC	personal computing
PCA	Prioritized Contention Access
PCI	Peripheral Component Interconnect
PDA	personal digital assistant
PDU	Protocol Data Unit
PE	portable electronics
PER	Packet Error Rate
PHY	Physical (layer)
PLCP	Physical Layer Convergence Protocol
PLME	Physical Layer Management Entity
PMD	PHY Medium-Dependent
PMK	Pre-shared Master Key
PMK	Pair-wise Master Key
PPDU	PHY Protocol Data Unit
PPDU	PLCP Protocol Data Unit
ppm	Parts Per Million
PRBS	PseudoRandom Bit Stream
PSD	Power Spectral Density
PSDU	PLCP Service Data Unit

PT	Preamble Type
PTK	Pair-wise Temporal Key
PTKs	Pair-wise Temporal Keys
QAM	Quadrature Amplitude Modulation
QoS	Quality of Service
QPSK	Quadrature Phase Shift Keying
RCI	Radio Controller Interface
RF	Radio-Frequency
RM	Range Measurement
RSSI	Received Signal Strength Indicator
RTS	Request To Send
RX	Receive or Receiver
SAP	Service Access Point
SBD	Self-Beaconing Device
SDIO	Secure Digital Input/Output
SFC	Secure Frame Counter
SFN	Secure Frame Number
SIFS	Short Inter- Frame Spacing
SIG	Special Interest Group
SNR	Signal-to-Noise Ratio
SrcAddr	Source device Address
TDMA	Time Division Multiple Access
TDS	Time-Domain Spreading
TFC	Time–Frequency Code
TFI	Time–Frequency Interleaved
TFI2	Time–Frequency Interleave with two bands
TG	Transaction Group
TKID	Temporal Key Identifier
ToA	Time of Arrival
TWR	Two-Way Ranging
TX	Transmit or Transmitter
TXOP	Transmission Opportunity
UDA	Unused DRP Reservation Announcement
UDR	Unused DRP Reservation Response
URC	UWB Radio Controller
URCD	UWB Radio Controller Driver
URCI	UWB Radio Controller Interface
USB	Universal Serial Bus
USB-IF	USB Implementers' Forum
UWB	Ultra wideband

WCDMA	Wideband Code Division Multiple Access
WCTA	W-USB Channel Time Allocation
WHC	Wireless Host Controller
WHCD	Wireless Host Controller Driver
WHCI	Wireless Host Controller Interface
WLAN	Wireless Local Area Network
WLP	WiMedia Logical Link Control Protocol
WPAN	Wireless Personal Area Network
W-USB	Wireless USB
WWAN	Wireless Wide Area Network
ZP	Zero Postfix
ZPS	Zero-Padded suffix

Appendix B

WiMedia Alliance Membership

This compilation of WiMedia Alliance membership is from http://www.wimedia.org/en/about/ourmembers.asp?id=abt as of July 2008. This list shows the strong support for WiMedia UWB and the ecosystem it can generate to promote the proliferation of this technology.

Promoters and Board Members
Alereon, Inc.
Cambridge Silicon Radio, Plc
Intel Corporation
Nokia Corporation
NXP Semiconductors
Samsung Electronics Co., Ltd
Staccato Communications
Stonestreet One
Wisair Ltd

Contributors
Advantest Corporation
Agilent Technologies, Inc.
Allion Test Labs, Inc.
AMD
Analog Devices
Artimi Inc.
ASTRI (Hong Kong Applied Science and Technology Research Institute)
AT4 wireless

Broadcom Corporation
Buffalo, Inc.
Canon Inc.
Dell, Inc.
Ellisys
Ericsson AB
ESRD of CSIST
ETRI (Electronics and Telecommunications Research Institute)
Faraday Technology Corporation
FOCUS Enhancements, Inc.
France Telecom Group
General Atomics
Hitachi, Ltd
Institute for Infocomm Research
ITI Techmedia Limited
JMicron Technology Corporation
LeCroy Corporation
Marvell International, Ltd
Matsushita Electric Industrial Co. Ltd

WiMedia UWB: Technology of Choice for Wireless USB and Bluetooth   Ghobad Heidari
© 2008 John Wiley & Sons, Ltd

MCCI
Microsoft Corporation
MindTree Consulting Ltd
Murata Manufacturing Co., Ltd
National Technical Systems
NEC Electronics Corporation
Oki Electric Industry Co., Ltd
Olympus Communication Technology
  of America
Omron Corporation
Realtek Semiconductor Corp.
Renesas Technology Corporation
Ricoh Company, Ltd
Royal Philips Electronics
Siemens Home and Office
  Communication Devices
Sigma Designs, Inc.
Sony Corporation
Symbian Software Ltd
Synopsys Inc.
Taiyo Yuden Co., Ltd
TDK Corporation
Tektronix, Inc.
Telecommunication Metrology Center
  of MII
Telecommunications Technology
  Association (TTA)
Toshiba Corporation
TZero Technologies
Verizon Communications, Inc.
WiLinx Corporation
Wipro-NewLogic
WiQuest Communications

Adopters
ACON, Advanced Connectek, Inc.
Actiontec Electronics, Inc.
Alcor Micro, Corp.
Alpha Networks, Inc.
ALPS Electric Co., Ltd

Chipidea – Microelectronica, S.A.
ELECOM CO., Ltd
Elonics
Fuji Electric Device Technology Co.,
  Ltd
Fujitsu Limited
Hewlett-Packard Company
Hong Kong University of Science and
  Technology, Department of EEE
iAnywhere Solutions
Institute of Microelectronics of Chinese
  Academy of Sciences
Integrated System Solution Corp.
IntellaSys, TPL Group Enterprise
IOGEAR
Konica Minolta Technology, U.S.A.
KTwo Technology Solution Pvt. Ltd
LG Electronics, Inc.
LitePoint Corporation
LucidPORT
Mercedes-Benz Research &
  Development North America, Inc.
MIPS Technologies, Inc.
MITAC International Corporation
Nano LSI Inc.
Parrot SA
Quanta Computer, Inc.
Quintic Corporation
RIC in Yeungnam University
Samyoung Electronics Co., Ltd
Sanyo Electric Co., Ltd
Seiko Epson Corp.
Silex Technology America, Inc.
Sporton International Inc.
TES Electronic Solutions
Testronic Labs/PMTC n.v.
TUV Rheinland
UBeacon Technologies, Inc.
Universal Scientific Industrial Co., Inc.
VinChip Systems

Wispro Technology, Inc.
Y-E Data, Inc.
ZyXEL Communications Corporation

Supporters
7 layers AG
A-Logics
Aalborg University – Communications Dept.
Abocom Systems, Inc.
Adaptive Labs, Inc.
Advance Data Technology Corporation
Advanced Resources Corporation
Aftek Infosys Ltd
AirGATE Technologies, Inc.
Airmesh Communications Limited
Akita pref. R&D Center
Alinks Communications
Allied Telesis Labs
Arab Media Group
ARS Software GmbH
Artec Design
ASRock Inc.
Audio Visual Concepts, Inc.
AWAH System Inc
Azentek
Barco Silex
BBN Technologies
Beijing Institute of Technology
Beijing Union University
Benetel
Berner & Mattner GmbH
Billionton Systems, Inc.
Blackbird Technolgies
Blind Creek Associates
Brevisys Technologies
BroadVision Technologies Inc.
Cabrillo Technology Int'l Inc
Cambridge Consultants
Centre for Wireless Communications

CeraMicro Technology Corp.
CETECOM GmbH
chipMARKETING
Clasma Events, Inc.
Co-Active Communications Corp.
Comit Systems, Inc.
ComNets, Chair of Communication Networks, RWTH Aachen University
Compliance Certification Services
Comscient Group
Concentric Technology Solutions, Inc.
Conceptronic
Concrete Logic
Corad Technology Limited
Current Electronics, Inc.
Daido Steel Co,. Ltd
Danbisoft
Datang Microelectronics Technology Ltd
Delta Electronics, Inc.
Denali Software, Inc.
Dialogic
Digipos Systems Inc.
DisplayLink Corp.
Doubleh Technology Inc.
Eagle Software India Private Limited
ELECTRODRMZ Consultants
Electronics Testing Center, Taiwan
EMC&EMB LAB BUPT CHINA
Engineering Club at SFCC
Ensphere Solutions, Inc.
ENSTA
EPCOS
ETS Product Service (USA), Inc.
Euphony Tech
EXODUS S.A.
FCC
FDK Corporation
FirmLogix

Foxlink/Cheng Uei Precision Industry
   Co., Ltd
FrameNet Inc.
Frontline Test Equipment, Inc.
Fujitsu Siemens Computers GmbH
Future Wireless Technologies
General Dynamics Canada
Global Edge Software
Globalintech Inc.
Golden Ocean International
   Techno-business Co., Ltd
Guardian Technologies Pty Ltd
GWT-TUD GmbH
Hakuto Co., Ltd
Hana Micron Inc.
Harman/Becker Automotive Systems
   GmbH
HauteSpot Networks Corporation
HCL Technologies Ltd
HEADESIGNS Inc.
HelloSoft, Inc.
Highlander Technologies, Inc.
Himico Solutions, Inc.
Homer Technology, Inc
Hoppmann Audio Visual
Horner Networks, LLC
I/O Interconnected
Icron Technologies Corporation
IJAK
Impact Technology Pte Ltd
IMST GmbH
INCUBE TECH CO., LTD
Indian Institute of Science
IndusRAD
Infinite Data Storage Ltd
Innovative Logic Inc.
Innovative Semiconductors, Inc.
Insys Canada
Intelligent Solutions2000
Interactive Homes, Inc.

ITC Engineering Services, Inc.
J&S Telecoms Int.
J.D. Taylor Associates
J3 Engineering
Jabil Circuit
Jazzhipster Corporation
Joints Chips Technology Pvt Ltd
JOOHONG
Kanaden Corporation
Kasshku Information Technologies (P)
   Ltd
Kiyon
Kodak
Korea Testing Laboratory (KTL)
KTL Ltd
La Salle
Levitate Technologies, LLC.
Lite-On Technology Corp.
LittleAt Pte Ltd
Logic IP, Inc.
Macronix
Mango Technologies Private Limited
Mantaro Networks Inc.
Maxi World Technology Limited
Media Device Lab
MET Laboratories
MG Systems & Software, LLC
MITE Global Communications
   Systems S.A. de C.V.
MLWizard
MRL Technology
MTI
N Square Corporation
Nanjing University of P&T
National Institute of Information and
   Communications Technology
National Physical Laboratory
Netplan A/S
NetStreams
Newport Technologies

North Carolina State University
NTS Calgary
NTT MCL, Inc.
Oak Tree Wireless
Open Interface North America, Inc.
OPHIT
Origo Ecuador
Oxford Semiconductor
PeerLink Corp.
Pinpoint Technologies, Inc.
Pinyon Technologies
Plextek
Positive Edge ASICs Inc
Prancer Corporation
Progress Software
ProyectoCid.ES
Q3 Wireless Inc
Raritan Computer, Inc.
RF Integration Inc.
RFI Global Services Ltd
Rheinmetall
Richwave Technology Corporation
Rohde & Schwarz System &
    Communications Asia Pte Ltd
Scheelite Technologies LLC
Sena Technologies Inc.
SGS TAIWAN LTD
Shenzhen Futurelooks Automation
    System Co. Ltd
Shenzhen Hiconn Electronics Co., Ltd
Signia Technologies, Inc.
Silicon Technology Co., Ltd
SK Telecom
SkillsUnited
SNU
South China University of Technology
Southeast University
Southwest University of China
SouthWing
Spansion

SunriseCottage
Suzhou Industrial Park Administrative
    Committee
Symbol
System Level Solutions, Inc.
Tadlys
Tag Guard Ltd
Tanvana Technologies
Targus
Tata Elxsi Ltd
TCH Enterprises
Technische Universität München
Teckwerks Limited
TECOM CO., LTD
TecStar Company, MACNICA, Inc.
Tek Elements, Inc.
Teknosarus Embedded Systems
Tektron Micro Electronics
Thales Joint Systems
The Aerospace Corporation
The Boeing Company
The Ether Group
The Institute of Information Science
    Academia Sinica
The University of Reading, Signal
    Processing Lab
Tops Systems Corp
Touch360
Trellis Communications
TRLabs
Twinhead International Corp.
Twomir Co., Ltd
Tyco Electronics AMP K.K.
UniqueLab Inc.
UMEC
Universidad Autonoma Metropolitana
Universidad Publica de Navarra
University of Applied Science
University of Kaiserslautern
University of Maryland

University of Pretoria
UnWiredConnect Technologies
   (P) Ltd
USCredential
Vayu Image
Vector Analytics, Inc.
Vestel Electronics
VIA Technologies, Inc.
VIDEOCORP
Virage Logic
Virtual Wire Technologies
Visitech AS
Vistapex Technologies
VoEx International LLC
W&W Communications, Inc.
W5 Networks

WhizNets Inc.
Wi2Wi, Inc.
Winbond Electronics
winsinfotek Pvt. Ltd
Wionics Research - Realtek Group
Wireless & Mobile Solutions, Inc.
Wireless USB Blog
wiSight, Inc.
Wisme, Inc.
WUSB Research Team
XtremeRF
Yokogawa Electric Corporation
Zappa Iberica
ZDSL.com
Zero Virtual
Zoran Imaging Division

Index